MATH FOR THE REAL WORLD

BY JERRY ATEN

ILLUSTRATED BY PAM BOONE

COVER DESIGN BY JEFF VAN KANEGAN

PHOTO CREDIT: COREL CORPORATION

PUBLISHER
INSTRUCTIONAL FAIR • TS DENISON
GRAND RAPIDS, MICHIGAN 49544

ISBN: 1-56822-432-X
Math for the Real World

2400 Turner Avenue NW
Grand Rapids, Michigan 49544

TABLE OF CONTENTS

INTRODUCTION

Math is everywhere! And it is very much a part of the lives of middle school students, even though they may not realize it. The activities and teaching strategies found in *Math for the Real World* really do come from real world experiences. Content-rich problems show students how mathematics is much more than numbers and how it has applications that count in their everyday lives. Placing mathematics into simulations that are realistic will help students to see not only its relevance, but how it spills over into other subject areas they are dealing with as well. Students begin to experience mathematics as an integral part of their world.

Basic facts are definitely the cornerstone of mathematics, and students can always use practice in improving these skills. But the ideas in this book will take students beyond this level to the point where they begin to make connections between the problems presented and their application in the real world.

There are some really big numbers in some of the activities. Teaching students to relate to those numbers is a part of the number sense approach that is so much a part of this book. Middle school students are always concerned about money, and the money unit will give them "hands-on" experience in learning how to responsibly manage money. Handling a checking account, using a credit card, and learning about budgets are just a few of the topics covered.

Many assignments relate to statistics and their interpretation. Numbers that define the profile of America in the unit "We the People" will give students a more realistic picture of the United States. Assignments on population data, big city comparisons, crime rates, the importance of education, and age differences are some of the topics that dish up some food for thought to students about their futures and quality of life.

The unit on buying, selling, and trading offers problems that relate to an important aspect in the life of the emerging adolescent. Students learn about comparison shopping, ordering off a restaurant menu, how much to tip the waiter, buying a house, buying a new car, buying life insurance, and the joys and woes of the sports card trading business.

The sports page becomes the focus of a unit in which students analyze some of the daily deluge of data landed on us by the sportswriters of America. Team statistics, individual performance data, outlandish salaries, and paid attendance figures are among the numbers that attract many of us daily (and middle school students, too!) to the sports page.

The unit "Myself and Others" contains activities that focus on students themselves. Counting heartbeats, burning calories, analyzing weights, and surveying television viewers provide the focus for some of the activities in this unit. The different twist to this unit is that students gather their own data in some of the assignments, then provide their own analysis.

Most of the activities in this book are more than a single page in length. Allowing students to work in pairs is strongly suggested for the assignments that contain listings of data or large numbers. For such assignments, hand-held calculators should also be available.

To make the activities more realistic to students, many of the questions call for an interpretation or evaluation on the part of the student. Allow them to make their own choices and exercise their own opinions and values. It is all a part of making it more real to them.

Teacher Notes

MONEY, MONEY, MONEY

Middle school students are at the age when money suddenly becomes very important to them. Learning about the realities involved in managing money is what this unit is all about. While most students are aware of many of the prices of goods and services that are important to their everyday lives, their major expenses are paid for by their parents. But most will also have some sort of income through an after-school or weekend part-time job, or they are getting an allowance from their parents in exchange for performing a few odd jobs around the house.

A good way to open this unit is to entertain an open discussion of what your students currently do to get the available spending money they have. Allow exchange of ideas that also includes varying values placed on either jobs performed for other people or for their parents. Those students (if any) who have no source of income other than what their parents merely hand them on occasion should also be allowed to speak and voice their opinions.

Activities in this unit will take students all the way from exploring various denominations of money to an examination of the federal budget. They will create a personal budget and learn about checking accounts and credit cards. The math involved in completing the activities will highlight the importance of having a good background in math to be able to deal with the realities of money in the real world.

Face Value

This opening activity gives students a brief overview of money including its early history and various places where it is minted here in the United States. Students will know many of the portraits used on denominations of money currently in use. But they may need to do a bit of research to properly identify that person's "claim to fame." The final task in the activity will no doubt require some research to properly identify those famous persons whose portraits appeared on denominations of money no longer in circulation.

Megabuck$

The task of this assignment is to use some "big number" math to arrive at the amount of money that is currently in circulation for each of the denominations of money used. The final BIG, BIG number is the total amount of money in circulation during a recent year. Students will not only get practice in dealing with adding and multiplying big numbers, but they will also practice reading those numbers.

Foreign Currency

Prior to assigning this activity, ask students if any of them know the names for any currencies used in countries other than the United States. For those who can identify other currencies, probe further about possible values compared to the U.S. dollar. Use the final task in the activity to expand this concept of comparing foreign money with the U.S. dollar.

Big Budgets—Big Debts!

Students learn about the huge debts that have been accrued by the federal government since it first began spending more than it collected in 1969. The concept of how really large the current deficit is becomes realistic to them when they calculate how much money it would cost every man, woman, and child in the United States if we paid off that debt all at once! You can find the current debt (assigned for students to find) in a current almanac.

Where the Money Comes From . . . Where It *Goes*

This activity goes in two different directions in introducing students to the federal budget. First, students study the various sources of income collected by the federal government for use in providing the goods and services demanded by American citizens and taxpayers. Practice in the mathematics of very large numbers is the main goal here, but this should also be accompanied by a discussion of the constant struggle the government has to find sources of income to pay for the many services it provides. Second, students create a circle graph on their own of the various ways (and proportions) the government spends our money for those services.

Living Within My Means

Place students in pairs for this activity both to stimulate interaction and to give them a chance to compare their own numbers with those of a partner. The whole idea is to arrive at a realistic "family budget" for a family of four for one month. Since students may not be familiar with realistic numbers for some of the typical expenses (and may in some cases not get much help from home), it might be a good idea to share some realistic expenses for those areas where they may be experiencing problems. The big benefit of this activity will be the realism of the expenses involved in a family budget when each partnership shares its findings and compares those numbers with other partner groups within the class.

Checking It Out!

The objective of this activity is to give children experience and practice on how a checking account actually works. A real-life simulation of a typical family's income and expenses is provided. These are handled by the students themselves through a checking account. It is suggested that you assign the activity individually to impact the responsibility factor attached to having a checking account. Students learn about the essentials involved in writing a check plus the importance of keeping accurate records with the check register. Blank checks are provided. You will need to reproduce enough so that each student has 11 checks. Students will need to number and complete the checks. Not included in the activity, but certainly an appropriate extension, would be to show students how to reconcile a monthly bank statement using the numbers provided by the activity.

Plastic Peril

Surfacing throughout this activity is an alert to students of the dangers of the overuse of credit cards. Students are led to the conclusion that using credit cards with reckless abandon can lead to financial disaster. The buying spree, often typical of first-time credit card holders, is taken to a sad but realistic ending with the purchases of Calvin and Theresa Rigdon. As students unravel the problems involved with their overspending, they become keenly aware of the financial responsibilities and sometimes unfortunate hardships that parallel excessive credit card spending. It is suggested that you conclude this activity on a more positive note by highlighting the convenience and popularity of credit cards in the United States and all over the world. But responsible use should be primarily controlled by a family's income, which is the bottom line.

Name ______________________________

Face Value

During the Middle Ages when trade developed among European nations, gold was the popular medium for trade. But gold was heavy and difficult to carry, and the European cities and roads were dangerous places to carry large amounts of gold. As a result, goldsmiths began issuing notes that were promises to pay gold to the person who was carrying the note. They were in effect the first forms of paper money. The idea behind using coins that all looked alike as a medium for trade dates as far back as the seventh century B.C. in the city-states of ancient Greece.

Today, money is issued by governments throughout the world. Here in the United States, money is made by the Department of the Treasury and the U.S. Mint. This arm of the federal government is responsible for making all U.S. coins. The U.S. Mint's headquarters is in Washington, D.C. Local branches that produce coins are located in Philadelphia, Pennsylvania; Denver, Colorado; San Francisco, California; and West Point, New York. It also keeps safe the Treasury Department's stored gold and silver at a guarded vault in Fort Knox, Kentucky.

The Bureau of Engraving and Printing is another division of the Treasury Department and is responsible for all of the designs and printing of all U.S. paper money. It is also located in Washington, D.C.

The U.S. Mint coins dimes, quarters, half dollars, and dollar coins that look like they are made from silver, but in reality are a mixture of metals that includes silver, copper, and nickel. Pennies are made from copper and nickels are made from nickel. The U.S. Mint occasionally makes special coins honoring famous people and events. The Dwight D. Eisenhower $1 coin and the Susan B. Anthony $1 coin are examples of special issue coins.

The more traditional coins and bills should be familiar to all of us because we use them so often in our everyday lives. But how familiar are you with the portraits of the famous Americans that are included on these coins and bills?

On the accompanying page are the various denominations that are currently in circulation. How many do you know without looking elsewhere? Jot down the value and the name of the famous American in the space provided. Next to his name identify his claim to fame. Then do a little research on your own to discover the names of those with which you are not familiar.

COIN/CURRENCY	VALUE	FAMOUS AMERICAN	CLAIM TO FAME
penny	______	____________________	________________
nickel	______	____________________	________________
dime	______	____________________	________________
quarter	______	____________________	________________
dollar bill	______	____________________	________________
two-dollar bill	______	____________________	________________
five-dollar bill	______	____________________	________________
ten-dollar bill	______	____________________	________________
twenty-dollar bill	______	____________________	________________
fifty-dollar bill	______	____________________	________________
hundred-dollar bill	______	____________________	________________

Bills larger than $100 were made prior to 1969. Find out whose portrait was on each of the following larger denominations.

$500 bill ______________________________________

$1,000 bill ______________________________________

$10,000 bill ______________________________________

$100,000 bill ______________________________________

Name ____________________

Megabuck$

The United States Treasury Department is responsible for authorizing every last dollar that finds its way into circulation. It is also accountable for knowing precisely how much money is in circulation at any given point in time.

Below is a recent "update" of the number of bills in circulation. Your task is to calculate the amount of money in circulation for each denomination; then add them all together, and present a grand total that tells how much money was in circulation in the United States on this particular day.

DENOMINATION	NUMBER OF BILLS	AMOUNT OF MONEY
$1 bills	5,674,996,280	____________
$2 bills	486,955,360	____________
$5 bills	1,334,489,900	____________
$10 bills	1,320,505,628	____________
$20 bills	3,753,078,824	____________
$50 bills	820,922,986	____________
$100 bills	2,075,392,105	____________

TOTAL AMOUNT OF MONEY IN CIRCULATION— ____________

1. Which denomination had the greatest number of bills in circulation? ____________
2. Which denomination had the fewest number of bills in circulation? ____________
3. Which denomination of bills represented the greatest amount of money currently in circulation on the day these statistics were compiled? ____________

Foreign Currency

The money a country uses as its medium of exchange is called its *currency.* Here in the United States the dollar is the standard of our currency. All money that is minted or printed is based on either a fraction of a dollar (coins of value less than a dollar) or a multiple of a dollar (all bills carrying a value in excess of a dollar). It would be really convenient for us if the rest of the countries of the world used the dollar, too. But they do not.

Each country has its own currency and a name for that currency. The values attached to those various denominations are based on a number of factors. Even currencies bearing the same name often have different values.

The list in column one on the following page contains the names used for various currencies used around the world. The second list includes the names of the countries that use those currencies. Your task is to match them by letter in the blank space provided. Remember that some names of currency are used by more than one country. All countries will not be used.

Investigate one of the currencies in the list. Find out where it is used and its current value. How much would it be worth when exchanged for U.S. dollars? Share your findings with other members of your class.

Name ______________________

NAME OF CURRENCY	
1. ________	cruzeiro
2. ________	rand
3. ________	schilling
4. ________	krone
5. ________	dollar
6. ________	guilder
7. ________	bolivar
8. ________	peso
9. ________	peseta
10. ________	ruble
11. ________	rupee
12. ________	yuan
13. ________	koruna
14. ________	krona
15. ________	dinar
16. ________	drachma
17. ________	pound
18. ________	riyal
19. ________	yen
20. ________	zloty
21. ________	dirham
22. ________	franc
23. ________	lira
24. ________	mark

COUNTRY

A. China
B. Egypt
C. South Africa
D. Saudi Arabia
E. Russia
F. Denmark
G. Italy
H. Germany
I. Spain
J. Poland
K. India
L. Austria
M. Japan
N. Venezuela
O. Australia
P. Greece
Q. Netherlands
R. Czech Republic
S. Slovak Republic
T. Iceland
U. Brazil
V. Algeria
W. Morocco
X. Argentina
Y. Jordan
Z. Kuwait
AA. United Arab Emirates
BB. Cambodia
CC. Canada
DD. Norway
EE. Turkey
FF. Chile
GG. United States
HH. Pakistan
II. Belgium
JJ. New Zealand
KK. Lebanon
LL. Colombia
MM. Great Britain
NN. France
OO. Mexico

Big Budgets—Big Debts!

Families have to have budgets. Businesses have to have budgets. Governments have to have budgets, too. Budgets are really nothing more than game plans for keeping track of money that is spent and money that is received. The most important rule of any successful budget is no more money can be spent than is received.

Perhaps the greatest example of how NOT to run a successful business with a balanced budget is the United States Government, which has not had a balanced budget since 1969! Below are some examples of recent years in which the U.S. Government spent more money than it collected through taxes. Look at each year and calculate the deficit that was the result. Express your answer for each year in billions of dollars. Remember, there are one-thousand thousands in a million, one-thousand millions in a billion, and one-thousand billions in a trillion!

YEAR	$ TAKEN IN	$ SPENT	DEFICIT
1991	$1.031 trillion	$1.252 trillion	________
1992	$1.092 trillion	$1.382 trillion	________
1993	$1.153 trillion	$1.408 trillion	________
1994	$1.249 trillion	$1.484 trillion	________

1. In which of the years above was there the greatest difference between the amount of money received and the amount of money spent by the U.S. Government?

   ____

2. In which year did the U.S. Government come closest to a balanced budget?

   ____

Name ______________________________

3. In the four years that are presented in the chart, how much more money did the government spend than it received? ____________________

4. In a recent year the federal debt had escalated to $4.0646 trillion! If the population of the United States were 256,500,000, what was the debt per capita? Use the space below to calculate your answer. Then round your answer off to the nearest dollar and place it on the line. ________________________

5. Do a little research on your own and find out the amount of the current debt owed by the United States Government and place that figure on the line below. Write the entire number in digits. ______________________________

6. Do a little more research and find out how many people there currently are in the United States. You can find this information in an almanac. Divide the current national debt (from problem #4) by the current population to find out the *per capita debt* (which is the way governments measure the extent of their debt). Place your figure on the line below.

____

7. Has the debt increased or decreased from the calculation you made from an earlier year in problem #4 to the calculation you made for the current year in problem #6?

____

Name ____________________

Where the Money Comes From . . . Where It Goes

Those people who concern themselves with budgets must be constantly aware of these two issues: 1. Where the money will come from, and 2. How the money will be spent. Whether it is a simple family budget, the budget of a small business or giant corporation, or even the budget of an entire nation, the two major concerns remain the same: income versus expense.

Typically, those persons or departments or agencies that are expected to live within the fiscal limits of the budget are asked to predict how much money they will probably spend within a defined period of time. Those asked to prepare such numbers often make generous predictions of their needs, because they want to make certain that they have enough money.

The next step is to determine the available sources of income that will be needed to pay the needed expenses. Typically, those charged with preparing such predictions are often faced with the dilemma of not seeing enough income to offset the predicted expenses.

When this happens, one of two events must occur. Either the income must somehow be increased to match the predicted expenses, or the predicted expenses must be decreased to match the expected income. Oftentimes, the final budget is a compromise between the two. Regardless of the final outcome, the most important factor is to make certain that the expenses do not exceed the income!

Below is a graph that portrays the sources the U.S. Government used in a recent year to get the revenue to provide for all of the goods and services that were needed by the citizens of the United States.

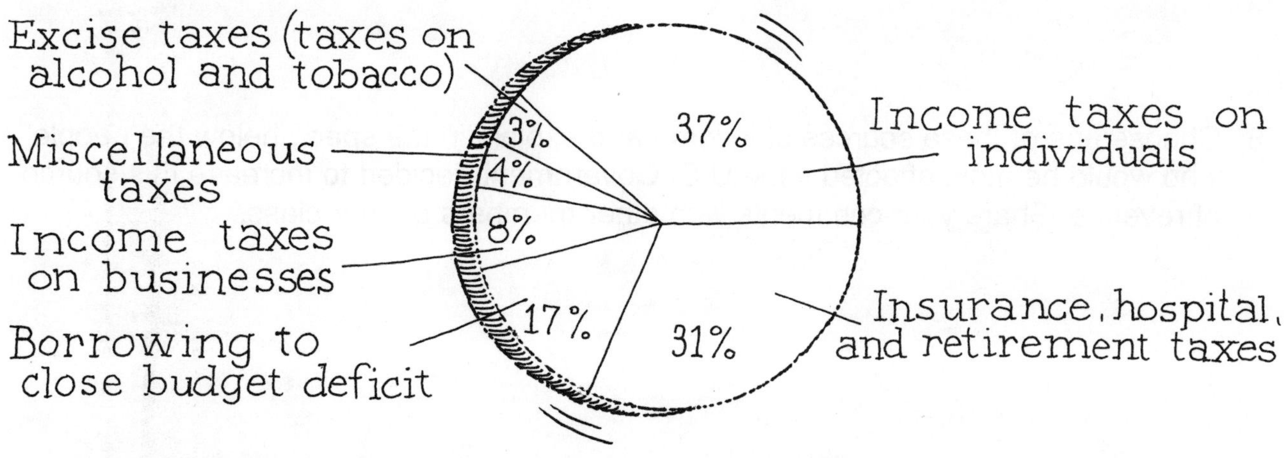

1. What was the government's greatest source of revenue? ____________________

2. What source of revenue amounted to 31 percent of total income? ____________

3. What percent of the revenue came from income taxes on businesses? __________

4. If the total amount of money taken in by the government were $1.249 trillion, calculate how much money was received from each source. Round off to the nearest billion dollars.

Income Tax on Individuals ______________________________

Insurance, Hospital, and Retirement Taxes ______________________________

Borrowing to Close Budget Deficit ______________________________

Income Taxes on Businesses ______________________________

Miscellaneous Taxes ______________________________

Excise Tax on Tobacco, Alcohol, Etc. ______________________________

5. Add together your calculations for all of the various sources of revenue. Does your total come close to $1.249 trillion? It should! If your calculations were far off, explain where you think you went wrong.

6. Choose one of these sources of income and explain in the space below the people who would be most affected if the U.S. Government decided to increase this source of revenue. Share your comments with other members of your class.

The U.S. Government spent $1.484 trillion during the same year that it brought in $1.249 trillion. This led to a deficit of $235 billion. Such deficits are not uncommon with the U.S. Government, which is trying to stretch tax dollars to cover the protection, welfare programs, and other services needed by Americans. Often the supply of money runs out before all of the goods and services are covered. When that happens, an unbalanced budget occurs and the government is forced to borrow money to make up the difference.

Below is an accounting of how the U.S. Government spent its money during the year described above.

46% —Social Security, Medicare, Medicaid, and Government Pensions
18% —Defense
15% —Payments of Money Directly to States and to Local Governments
14% —Interest on the Debt (the money borrowed to lessen the budget deficit)
6% —The Cost of Running the Federal Government
1% —Insurance on Savings Deposits

Portray these percentages in the form of a circle graph similar to the graph on the first page of this activity that graphically showed you the various income sources used by the U.S. Government. (Calculate the percentage of the entire graph of 360° that should be occupied by each section of the pie. Then mark off that section using a protractor and properly label it.)

Name ____________________

Living Within My Means

Families need to have budgets too! Just as our federal government can get deeply into debt by continuing to spend more than it takes in, a family can develop financial woes by overspending. The secret to the financial success of any family is really no secret at all. It just comes down to a practice of "living within that family's means." That means learning to live with a family budget.

Preparing a family budget can be an almost impossible task if the amount of money that will be coming in during a given period of time is a mystery. However, if the amount of income *is* known, preparing the budget becomes a matter of planning expenses that do not exceed the planned income.

Since the preparation of most family budgets involves an interaction of people, you will need a partner for this activity. You and your partner first need to brainstorm what you feel will be the various expenses needed to maintain a reasonably comfortable and safe standard of living for a family of four that includes a husband, a wife, a 15-year-old son, and a 12-year-old daughter. Since most expense bills are issued on a monthly basis, assume that this is a monthly budget of both expenses and income.

To make this more realistic, assume that this family lives in the same city or town where your school is located. In the space below, list the various kinds of expenses you think would be incurred by this family of four. Be certain to include all of the necessary costs you can brainstorm together. Do not forget to include some kind of a savings plan!

Next, you need to provide some estimates for what you think these various expenses will cost. To make the numbers more realistic, talk with your parents about what they feel some of these various costs should be for one month. Record their figures and share your numbers with your partner. If there is a difference between the two of you on a projected expense, simply add the two together, and divide by two. That will give you an "average" expense which you can use for your budget. List your categories and projected expenses below.

Now you need to total all the expenses. Record your answer here.__________

Remember that this total represents expenses for just one month. To find out a yearly total, multiply by 12 and record your answer here.______________

To live within the budget you have just prepared would require an annual income of at least that much money. Is this figure realistic for your area?
What kinds of jobs in your city or area have annual salaries that would be in line with this kind of living?

____

____

You and your partner need to classify your budget using the following criteria:

Is it a "bare bones" budget? Perhaps it is just "average living." Or does it deserve the rank of "high roller living"? Circle your choice and compare your budget with those of other partner teams.

Name ______________________

Checking It Out!

There are a couple of reasons most adults prefer to use checks over cash to pay their bills. Perhaps the most important reason is that they do not have to carry a lot of cash. But they also have a very accurate record of how much money they really have and how much they have spent paying their bills. The *check register* keeps a running total of what is placed into the checking account at the bank as well as a running total of the day-to-day checks that are written against the account. If there is ever a question about whether or not a bill has been paid, the check register would serve as an excellent reference.

Checking accounts thus offer some real conveniences to the user. How it all works is magically controlled through the modern technology of banks. Basically, it works this way. When a person writes a check to pay a bill, the person writing the check is promising to pay the recipient the amount that is stated on the check. The next step is to place that check in the possession of the recipient. This is often conveniently done through the simplicity of the U.S. Mail.

The recipient then takes the check to his or her own bank. There it is either cashed or the value (in dollars) written on the check is placed into the check recipient's account. That bank then sends the check back to the bank of the check writer. The amount of the check is deducted from the check writer's account. The check recipient's bank is then reimbursed in the amount of the check.

For banks to be paid for services such as this, they usually charge a nominal monthly *service charge* for checks that are written by those who have accounts there.

It is very important for those people who write checks to keep very accurate records of the checks they write. Otherwise, they might write checks that total more money than they have in the bank. When this happens, the checking account is said to be *overdrawn.* If this happens to a check writer on an infrequent basis, the bank will usually go ahead and honor the check, then call the check writer and ask that sufficient funds be placed into the account immediately to cover the check.

However, if this practice continues on a regular basis, the check writer will not only be in serious trouble with the bank, but will also be in trouble with the law! There are people currently occupying jail cells all over America because they were found guilty of engaging in the practice of writing *bogus,* or bad, checks!

This activity you are about to become involved with requires you to go through the steps involved in paying bills by check (actually writing the check) as well as keeping an accurate accounting of the checks you write.

Janet and Tim Wiggins have a checking account at the First State Bank. They deposit their paychecks into the account each payday; then they write checks against the account to pay all their bills and obtain cash for the "daily" expenses of living. Look closely at each transaction before you act. If it is a bill that needs to be paid, write a check just like the first entry for the month that is done below for you. Be certain to jot down the correct date. Make certain you spell correctly the name of the company or person being paid by the check. Write down the amount on the check *both* ways you see in the example. Be certain you sign the name of Janet Wiggins (who pays all the family bills). Many people also jot down the purpose for writing the check on the "For" or "Memo" line in the lower left-hand corner of the check.

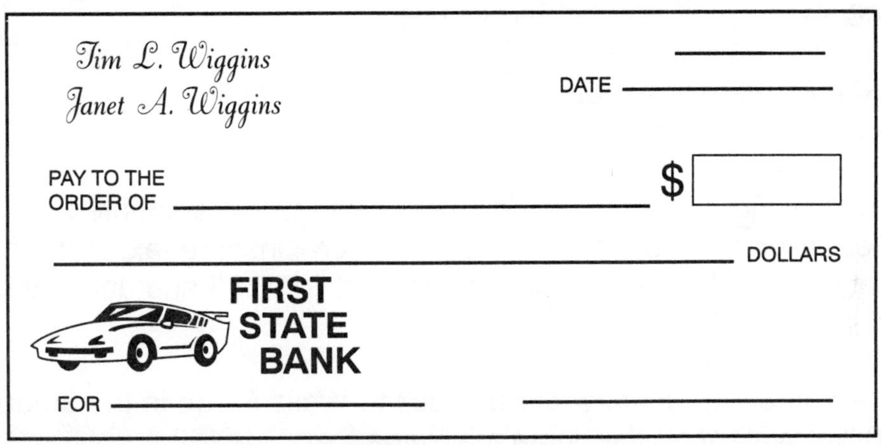

You should next go to the check register provided you and jot down the same information. This is done so that the check writer will have his or her own accounting of the check written after the check has been sent to the recipient. The final step is to subtract accurately the amount of the check from the balance. When you find a transaction that involves either Janet's or Tim's depositing a paycheck into their account, be certain to record this in the check register too. Those amounts should be added to the account balance. Carefully complete each of the following transactions:

— Opening balance of the account on June 3 is $493.61.

— That day (June 3) Janet picked up some groceries on the way home from work at Smitty's. She paid by check the total amount of $28.08.

— On June 4, Janet paid Air Touch Paging $32.33, the amount charged to Tim for their paging service for the months of June, July, and August.

— She also wrote a check that evening to Magpie's Pizza for a pizza that served as dinner for her family. The pizza cost $14.95, there was a $1.50 delivery charge, and Janet gave the delivery boy a $2.00 tip (all included in the amount of the check she wrote).

— The next day (June 5) Janet deposited part of her paycheck in the bank to help restore the family checking account. Her "take-home" check for two weeks' work after all the deductions amounted to $814.66. She kept $74.66 and deposited the rest into the checking account.

— That evening she wrote a check to Chevron in the amount of $76.15 for Tim's monthly gasoline credit card charges.

— The next day (June 6) she took Katie for her monthly orthodontal appointment with Dr. Augat. She and Tim have agreed to a monthly payment of $95 until the entire bill for Katie's braces has been paid. While there, Janet wrote a check to Dr. Augat for the month of June.

— On June 7, Janet wrote a check to the Building Fund at the Church in the Foothills for $40.00.

— That same afternoon she wrote a check to Shell Oil for $62.27 to cover her credit card charges for gasoline for the past month.

— On June 9, Tim brought home his paycheck, which amounted to $1,326.89, once all the deductions had been taken out for federal income tax, state income tax, social security tax, and health insurance. He kept $126.89 and deposited the rest into the checking account.

— On the same day, Janet wrote a check to Wells Fargo in the amount of $501.40 for the monthly payment on the home mortgage loan.

— She also paid off the water bill to Canada Hills Water Company for $25.16 with the next check in the family checkbook.

— She paid Catalina Power Company $147.88 for 34 days of electricity, her family's bill for last month's use.

— On the next morning (June 10), Janet ran her ATM card through a walk-up machine and withdrew $60.00.

— She bought groceries that morning at Fry's and paid the total bill by check, which was $75.16.

What is the current account balance Tim and Janet have in their checking account? _____

Jim L. Wiggins
Janet A. Wiggins

DATE ____________

PAY TO THE ORDER OF ______________________ $ [   ]

______________________ DOLLARS

FIRST STATE BANK

FOR ____________ ____________

Jim L. Wiggins
Janet A. Wiggins

DATE ____________

PAY TO THE ORDER OF ______________________ $ [   ]

______________________ DOLLARS

FIRST STATE BANK

FOR ____________ ____________

Jim L. Wiggins
Janet A. Wiggins

DATE ____________

PAY TO THE ORDER OF ______________________ $ [   ]

______________________ DOLLARS

FIRST STATE BANK

FOR ____________ ____________

Jim L. Wiggins
Janet A. Wiggins

DATE ____________

PAY TO THE ORDER OF ______________________ $ [   ]

______________________ DOLLARS

FIRST STATE BANK

FOR ____________ ____________

Jim L. Wiggins
Janet A. Wiggins

DATE ____________

PAY TO THE ORDER OF ______________________ $ [   ]

______________________ DOLLARS

FIRST STATE BANK

FOR ____________ ____________

Jim L. Wiggins
Janet A. Wiggins

DATE ____________

PAY TO THE ORDER OF ______________________ $ [   ]

______________________ DOLLARS

FIRST STATE BANK

FOR ____________ ____________

Check Register

Check No.	Date	Description of Transaction	Payment	Deposit	Balance

Name ______________________________

Plastic Peril

Just as Americans love the convenience of checking accounts, so do they also have a passion for credit cards. It is all so simple! All you do is walk into the store, decide what you want to buy, take your merchandise to the counter, present your credit card, and a few minutes later you walk away with your new purchase without ever paying any money. How easy can it get! To make it even better, you can even insert your credit card into an automatic teller machine, tell it how much money you want, and in a matter of a few seconds, you have your instant cash!

Credit cards *do* provide convenience to consumers because they make it unnecessary for the cardholder to carry a lot of cash. The process is simple . . . and IT WORKS! Wow, does it ever work! Americans are in debt billions of dollars because of it.

Sometimes credit card holders get into excessive debt because they do not control their use of the credit cards. They do not use the card just for convenience; they become addicted to buying as much "stuff" as their credit limit will allow. Then they worry about how they will pay for the charges they have incurred at a later date. This is not the way people should organize their financial lives.

Credit card companies rely on this "overspending" by Americans to make their money. Most credit card companies impose a monthly interest rate of at least 1.5% or 18% per year. Many companies stretch the rate even beyond that. The card holder who extends his or her credit margin to the limit, then pays only the monthly minimum, winds up paying a significant amount of money for the use of the credit.

For example, Calvin has a MasterCard which he got for "free" (no annual fee) through his local bank, the Happy Valley Bank. Calvin's card carries a $5,000 credit limit that was based on his current credit record. They gave Calvin's wife Theresa a card, too, but both are under the same account. Since receiving their new credit cards, both Calvin and Theresa have been quite busy keeping the plastic warm as they have made a number of purchases during the past month. It seemed like so much fun! Buying things they could not otherwise afford brought great happiness into their lives. But alas, the monthly bill has arrived. A copy of that statement appears on the next page.

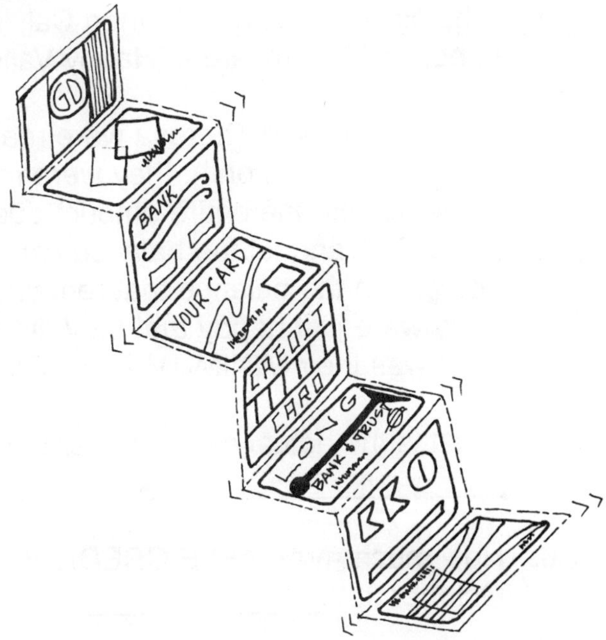

HAPPY VALLEY BANK MASTERCARD

Acct. No.	Credit Line	Avail. Credit	Bill Date	Min. Pymt.	Due date
5246 8002 0421 0628	5,000	$1,743.40	10/31/96	$97.70	11/18/96

DATE	DESCRIPTION	TOTAL
10/06	LAZ-Y Boy	$625.46
10/06	Toys R Us	48.13
10/08	Pet Smart	22.98
10/08	Monument Camera	312.42
10/09	Home Depot	492.16
10/12	Champs	48.14
10/15	Target	29.47
10/15	Brake Masters	15.15
10/16	Circle K	15.00
10/20	Circuit City	625.18
10/20	Egghead Software	49.85
10/24	California Design	814.91
10/28	Nature Company	52.08
10/30	Desert Landscaping	105.67
		$3,256.60

Prev. Bal.	Payments	Credits	New Purchases & Advances	New Bal.
0.00	0.00	0.00	3,256.60	3,256.60

For customer service call 1-800-904-8357

Look carefully at Calvin's bill and answer the following questions.

1. When Calvin and Theresa got their bill in the mail, they could not believe their eyes! Wow! What was the total amount of their purchases since they started their new habit of "buying with plastic"? ____________

2. What was the BILL DATE noted on their monthly statement? ____________________

3. What is the DUE DATE when Calvin and Theresa must have their money in the hands of MasterCard at Happy Valley Bank? ____________

4. While Calvin and Theresa were really shocked that they had bought so many things during the past month, they were relieved to see that they did not really have to come up with the entire amount due on their bill. They could make the MINIMUM PAYMENT and let the balance carry over. They knew they would be charged interest on the balance that remained, but that did not matter to Calvin and Theresa. They were just happy that they did not have to come up with the entire balance. So, what was their MINIMUM PAYMENT? __________

5. What is the actual card number of Calvin and Theresa's MasterCard?

____

6. How much AVAILABLE CREDIT do the Rigdons have on their new account as of October 30? __________________

7. Explain how Happy Valley Bank arrived at this amount of remaining credit available to Calvin and Theresa. ____

____

8. What should the Rigdons do if their new card should become lost or stolen?

____

9. What percent of the total balance due was the MINIMUM PAYMENT which Calvin and Theresa could make? ______________

As the Rigdons entered the month of November, they began to think ahead toward some early Christmas shopping. Theresa carefully planned her list ahead of time and then methodically checked off each gift as she made the purchase. Of course, she used her new credit card "for convenience" to buy most of the gifts. She was very careful to save copies of each receipt for two reasons. The first was in case the gift she bought for someone was not the right size or in some other way needed to be returned. Theresa also saved the receipts so she would have a running total of all her purchases.

But before Calvin and Theresa even started their Christmas shopping, they decided to buy a long overdue new refrigerator "to reward themselves" for years of hard work and "doing without many of the finer conveniences of good living." They found a great buy on a 21-cubic-foot, frost-free Whirlpool with a factory installed ice-maker for only $799! They even had their choice of colors. They just could not pass it up. With tax of $55.93, the total bill came to $854.93. Of course, by now we would expect them to use their credit card for such a purchase. And they did.

Now . . . on to the mall . . . and some early Christmas shopping!

10. Theresa and Calvin spent the first Saturday in November charging the following gifts: a beautiful piece of pottery at Southwest Interiors for Calvin's mom and dad for $145.18, a lamp for Theresa's mother at Desert Digs for $93.17, a small pull cart for Theresa's dad at Nevada Bob's for $65.12, toys for their daughter Marie at Toys R Us for $145.15, a Super Nintendo for their son Tyler at Circuit City for $167.19, and a great clock for Theresa's sister at Furniture Creations for $59.22.

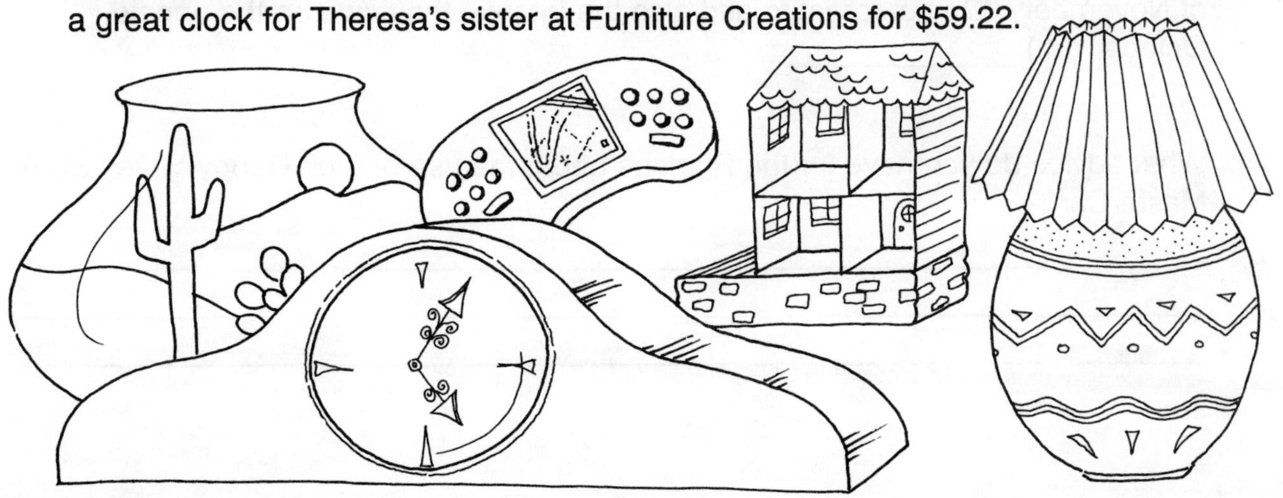

Two days later Calvin bought Theresa a nifty cocktail ring at The Diamond Mine for $284.56. That same evening Theresa went to Van's Golf Pro Shop to buy Calvin a new set of woods he had been wanting. The total bill came to $314.55. When she went to pay for the woods with her Happy Valley Bank MasterCard, the authorization came back to the clerk with the message that the "available line of credit extended to this card holder would be exceeded by this purchase."

This sadly meant that Theresa could not buy Calvin the woods she wanted to buy him for Christmas. She tearfully went home and totaled up the receipts she had so carefully preserved. She was once again shocked! What was the total of the purchases she and Calvin had charged during the month of November (excluding the golf clubs, which she did not get to charge)? ____________

11. Theresa and Calvin had a long talk about their two-month buying binge and decided that maybe it had not been such a good idea after all. They decided to pay off their bill completely before they started using the card again. Then they got their bill for November. Calvin then admitted buying Theresa the cocktail ring as a gift. First, how much interest would they pay at a rate of 18% per year (1.5% per month) on the unpaid balance they carried over from October? ____________

12. With their additional purchases in November, what would be the total amount on their November statement that would include the unpaid balance from October, the interest due on that unpaid balance, plus the new purchases made in November?

    ____

13. If the MINIMUM PAYMENT is 3% of the balance from last month plus 3% of the new purchases, what is the minimum payment the Rigdons can get by with for the month of November? (Do not forget to add also the interest they owe on the unpaid October bill.) ____________

14. What advice do you have for the Rigdons about the use of their Happy Valley Bank MasterCard?

    ____

    ____

    ____

Teacher Notes

WE THE PEOPLE

There are some really big numbers associated with population counts and the statistics that result from the analysis of these numbers! Students will have a number of opportunities to read and solve mathematical problems in this unit saturated with large numbers. Several of the activities focus on the 1990 Census conducted by the U.S. Bureau of the Census. Some of the questions asked by the Bureau served up some of the data used in this activity. Calculating percentages, rank ordering, graphing statistical data, reading maps, and interpreting data from maps are among the math skills required in the activities and strategies in this unit.

Everyone Counts!

Students calculate the population diversity of the 1990 U.S. Census. Discuss with them the reasons for conducting a census every decade and the importance of obtaining an accurate count. Encourage students to practice "rounding off" when dealing with large numbers to help give the numbers more significance and lasting meaning. It will be helpful to give a brief explanation of the steps used in rounding off, and several examples of practice will give students more confidence when they actually encounter the problems.

The Biggest Cities on Earth

Students work with large numbers calculating problems associated with the growth of some of the world's largest metropolitan areas. The basis of comparing one population to another focuses on the number of people *per square mile.* A worthwhile extension of the assignment is to have students do a bit of research in finding out the population density of their own city or town.

Big City, U.S.A.

The focus of this activity is the 1990 Census figures for America's ten largest cities. Students compare current statistics with the same-city populations of 1950 and determine which have grown and which have decreased in size. When members of the class exchange answers, discuss reasons students may feel certain cities have gained (or lost) in population.

Where We Live

Distribute to each student a copy of the map depicting the percentage change in population experienced by each state during the 1980s. Invite students to give their own suggestions on how to interpret the data presented. The activity also involves a graphing exercise on the various age levels of Americans. Allow students to exchange their bar graphs with other students to determine various possibilities of presentation.

The Graying of America

Before assigning the activity, be certain to discuss with students the definition of the term *median.* Students analyze reasons for some of the information that is gathered by the U.S. Census Bureau every ten years. The focus is the median age of Americans in each of our 50 states. Since Americans are living longer, there are a number of reasons to be concerned about where these older people are living. The activity becomes a real-world application of the significance of numbers because of the high interest in the data of various agencies and companies associated with the economy.

Equal Representation

Begin the lesson by sharing with students the historical background information that provided the motivation behind a bicameral system of lawmaking. Dating all the way back to the Constitutional Convention, the highly populated states wanted representation in the legislative branch of government to be based solely on population (i.e., the more people, the more lawmakers). The smaller states feared domination by the larger states and thus demanded that every state be allowed the same number of lawmakers. This impasse was the single most controversial issue in drafting the Constitution, and it eventually resulted in the bicameral system we have today that satisfied both sides.

Students again interpret data presented in a map showing the numbers of representatives allowed each state in Congress. Provide students with current almanacs that will allow them to research the statistics required to answer question #5.

Metro Crime

Distribute a copy of the fact sheet showing the various crime rates in some of America's major metropolitan areas. Be certain students clearly understand what the numbers mean. Review the definitions of *violent crimes* and *property crimes* found in the student introduction. The numbers given represent the number of crimes committed per thousand residents during the past calendar year. The overall ranking of each metropolitan area is a composite of the top crime rates.

As in other activities in this unit, stimulating students to think about possible reasons for the numbers being as they are in various cities is far more important than the numbers themselves.

Education Pays!

Students analyze data presented that shows the *mean income* for people with varying levels of educational achievement. The chart also includes a breakdown on the difference between the mean incomes for women and men at the various levels. Age is one of the variables that becomes a determining factor in what salary a person makes. The last page of the assignment has students calculating an entire career's earnings at the various levels of educational achievement. The assignment provides an excellent avenue to discuss at length student opinions on the value of an education.

Heating Our Homes

The census conducted every decade by the U.S. Bureau of the Census also asks Americans to indicate the source of energy used to heat their homes. Students are given the percentages and asked to calculate the number of households heated with each of the energy sources indicated. Students then portray their calculations by creating circle graphs.

Name ____________________

Everyone Counts!

Every ten years the United States Bureau of the Census conducts a population count. The census costs the U.S. government a lot of money, but it is important because of the information it reveals. One of the most important findings of the census is that it reveals trends where people are moving as well as where they may be leaving. The census also determines how many representatives in Congress each state shall have for the next ten years.

Such a count of people helps the national government to make plans to provide public services such as health care, highways, and new parks. It also helps state and local governments to decide on more local issues, such as whether to build more schools for children or homes for elderly people.

The census also reveals how old people are, what they do to earn a living, how much money they earn, how many children they have, and many other things about them. Private companies make use of this wealth of information in a variety of ways to help them plan for future products and their distribution.

The census also helps to find out the racial or ethnic background of Americans. The concept of race as used by the Census Bureau reflects self-identification. It does not denote any clear-cut scientific definition of biological stock. The data for race represent self-classification by people according to the race with which they most closely identify.

It should also be noted that the Hispanic population can be people "of any race" and will account for the fact that the chart below exceeds 100%. The chart contains the racial mix in the United States as determined by the 1990 Census.

White	**80.3%**
Black	**12.1%**
Hispanic	**9.0%**
Asian (including the Pacific Islands)	**2.9%**
American Indians (including Eskimo, or Aleut)	**.8%**

1. The 1990 Census included a final count of 248,709,873 people living in the United States. Based on this number, how many people (rounded off to the nearest thousand) indicated racial identification with the following groups?

   White ____________________

   Black ____________________

   Hispanic ____________________

   Asian ____________________

   American Indian ____________________

2. Create a pie graph in the circle below that will show the segment of the total population of the United States occupied by each racial group. Since the total numbers above exceed 100%, you will need to estimate the portion of the circle to be occupied by each racial group. To calculate the portion you should mark off in the circle for each group, multiply the percent by 360 (the number of degrees in a circle). Then mark off that portion of the circle using a protractor and label appropriately.

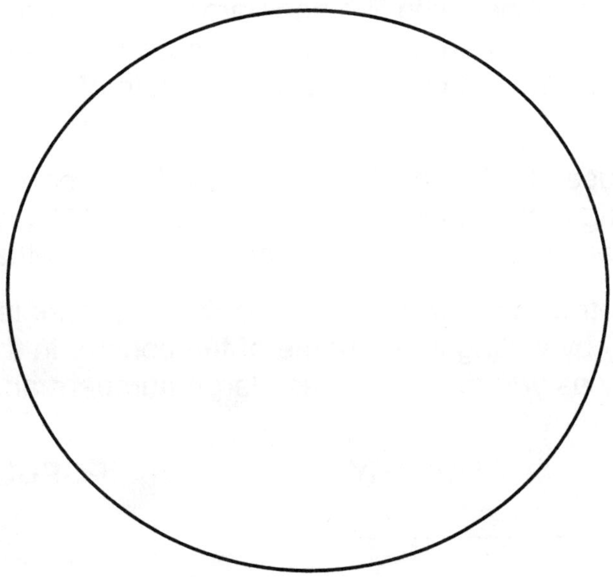

3. Question #7 on the Census Bureau questionnaire was stated in the following manner: "Is this person of Spanish/Hispanic origin?" Which of the following ethnic groups would you think accounted for more than 60 percent of Hispanics living in the United States? Circle your choice.

   Mexican Cuban Puerto Rican Other Hispanic

Name ______________________

The Biggest Cities on Earth

By the turn of the next century, over half of the people on earth who live in developed countries will live in cities. How we actually define a city allows for a wide latitude of comparison, depending on the definition used. The idea of 2,500 people being considered a city is perhaps appropriate for rural farming communities, but that definition would not be very meaningful when comparing it to life in a truly big city. To provide a more uniform gauge for comparing one large city with another, this activity will use the term only when making reference to areas with populations of more than 100,000 people.

People live in big cities for a variety of reasons. For many, it means more job opportunities. For others, cities offer more cultural opportunities. Big city life means excitement and convenience for those who live near large numbers of people. Goods and services are more accessible. Prices are much more competitive. People who live in large cities also have a much wider range of choices for making use of their free and leisure time.

Metropolitan area is a term used to define an area of very dense population. The term not only includes the people living in the city itself, but also the people living in areas around the city—in the suburban areas plus all of the unincorporated areas that are still close by.

Below are the ten largest metropolitan areas on earth. Your first task is to locate the metropolitan area by country by writing in the name of the country in the blank space provided. Then look carefully as you deal with these large numbers and answer the questions that follow.

CITY	COUNTRY	POPULATION
1. Tokyo-Yokohama	____________	27,245,000
2. Mexico City	____________	20,899,000
3. Sao Paulo	____________	18,701,000
4. Seoul	____________	16,792,000
5. New York City	____________	14,625,000
6. Osaka-Kobe-Kyoto	____________	13,872,000
7. Bombay	____________	12,101,000
8. Calcutta	____________	11,898,000
9. Rio de Janeiro	____________	11,688,000
10. Buenos Aires	____________	11,657,000

1. The world's metropolitan area with the largest population is spread over an area of 1,089 square miles. How many people are there per square mile in the Tokyo-Yokohama metropolitan area? ______________________

2. The Mexico City Metropolitan Area, on the other hand, is distributed over 522 square miles. How much more crowded is Mexico City than Tokyo-Yokohama? Use people per square mile to make your comparison. ______________________

3. The Bombay Metropolitan Area is the most densely populated of the areas in the top ten cities. Its crowded population is spread over only 95 square miles. How many people does Bombay have per square mile? ______________________

4. The most densely populated metropolitan area on earth is Hong Kong, which has a population of 5,693,000 people distributed over only 23 square miles! How does that population density compare with that of Bombay? Compare the two using people per square mile. ______________________

5. Of the metropolitan areas contained in the list on the first page of this activity, the "least crowded" is the Buenos Aires Greater Metropolitan Area, which has a population density of 21,790 people per square mile. How many square miles are there in the Buenos Aires Greater Metropolitan Area?

____

6. Which of these metropolitan areas is the more densely populated, Sao Paulo with an area of 457 square miles or Calcutta with an area of 209 square miles?

____

7. If the metropolitan area of the beautiful city of Seoul, Korea, has a population density of 49,101 people per square mile, how many square miles does the metropolitan area cover? ______________________

8. Of the metropolitan areas in the top ten, New York City sprawls over the largest area with 1,274 square miles. How many people does that mean live per square mile in New York City? ______________________

Big City, U.S.A.

The 1990 Census showed that almost 80 percent of Americans now live in cities and suburbs. The chart below shows the populations of the nation's ten largest cities in 1994. It also shows how those cities have gained or declined in population during the past four decades. Look carefully at the data presented and answer the questions that follow.

Rank	City	1950 Population	1994 Population
1	New York, New York	7,891,957	7,322,564
2	Los Angeles, California	1,970,358	3,845,557
3	Chicago, Illinois	3,620,962	2,783,726
4	Houston, Texas	596,163	1,629,902
5	Philadelphia, Pennsylvania	2,071,605	1,585,577
6	San Diego, California	334,387	1,110,554
7	Detroit, Michigan	1,849,568	1,027,974
8	Dallas, Texas	434,462	1,007,618
9	Phoenix, Arizona	106,818	983,403
10	San Antonio, Texas	408,442	935,393

1. Which six cities increased in population between 1950 and 1994?

   ____________________ ____________________

   ____________________ ____________________

   ____________________ ____________________

2. Which four cities showed a decrease in population during the same time period?

   ____________________ ____________________

   ____________________ ____________________

3. Which city experienced the greatest increase in the net gain in population for the time period indicated? ____________________

4. Which city experienced the greatest percentage increase in its population during the 40 years indicated by the chart? ____________________

5. Which city experienced the biggest loss in population during the time indicated?

   ____

6. Which city showed the greatest percentage of loss in population during the time period? ____________________

Name ____________________________

Where We Live

1. Among the many pieces of statistical data collected from the census was the fact that there were 121,239,418 males living in the United States and 127,470,455 females. What percent of the population is male and what percent is female? Round your answer to the nearest hundredth of a percent.

   Males represent _________% of the total U.S. Population.

   Females represent _________% of the total U.S. Population.

2. The census also found out the age of Americans. Below is a breakdown on numbers in various age categories.

Under 5 years	18,354,443
5 to 17 years	45,249,989
18 to 20 years	11,726,868
21 to 24 years	15,010,898
25 to 44 years	80,754,835
45 to 54 years	25,223,086
55 to 59 years	10,531,756
60 to 64 years	10,616,167
65 to 74 years	18,106,558
75 to 84 years	10,055,108
85 years and older	3,080,165

   A graph makes a much better presentation of this data than merely listing the numbers. In the space below create a bar graph that shows the above information. Estimate to the best of your ability the length of each of the bars; then fill in each bar with a black ink pen. Do not forget to label your graph to give the bars meaning.

3. The Census also showed those areas of the United States that are increasing in population and those areas that show little gain. Look at the breakdown of regions and note the populations taken during the 1980 Census compared to the 1990 Census.

Region	1980	1990
Northeast	49,135,000	50,809,000
Midwest	58,866,000	59,669,000
South	75,372,000	85,446,000
West	43,172,000	52,786,000

Which two regions showed the greatest growth between 1980 and 1990?

____

What was the percentage gain in population for each region between 1980 and 1990? Round to the nearest tenth of a percent.

Northeast __________

Midwest __________

South __________

West __________

4. Can you think of any reasons for the growth in the two regions that showed a significant increase in population between 1980 and 1990?

____

____

____

Look at the percentage change map that accompanies this assignment to answer the following questions:

5. Which five states showed the greatest percentage of increase in population between 1980 and 1990? List them in descending order, from #1 to #5.

#1. _______________

#2. _______________

#3. _______________

#4. _______________

#5. _______________

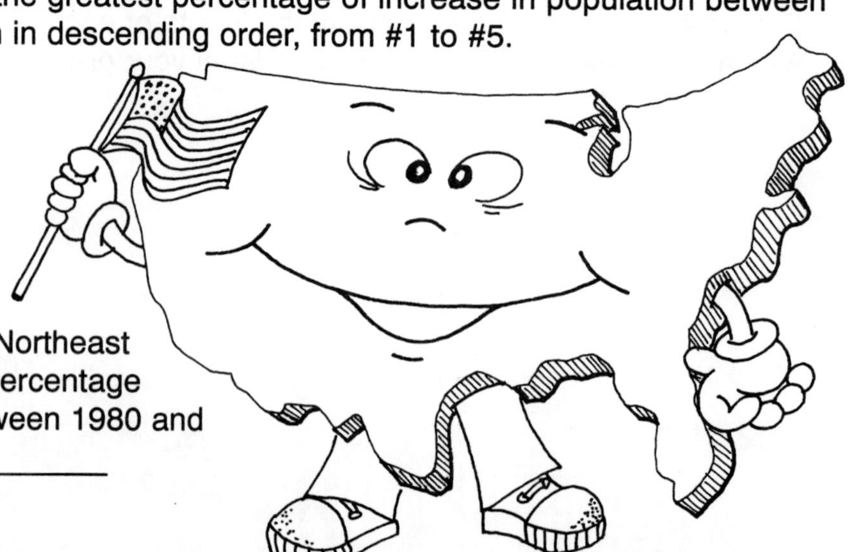

6. Which of the states in the Northeast experienced the greatest percentage increase in population between 1980 and 1990? ____________________

7. Which of the states in the Midwest showed the greatest percentage increase in population between 1980 and 1990? ________________

8. Which four states actually lost people between 1980 and 1990? List in descending order of percentage loss beginning with the state that experienced the greatest percentage loss.

   #1 ________________

   #2 ________________

   #3 ________________

   #4 ________________

9. Which of the states in the South showed the greatest percentage gain in population?

   ____

10. Which state considered as one of the Southern states actually lost people?

   ____

11. Which of the states in the West experienced the greatest percentage of increase in population between 1980 and 1990?

   ____

12. Which of the states in the West would you guess experienced the greatest actual gain in its number of people during the 1980s?

   ____

13. Which of the states in the West actually lost people during the 1980s?

   ____

14. Since the 1990 Census was taken, one of the states in the West has experienced a dramatic shift from a state that gained huge numbers of people during the 1980s to one that has large numbers of people leaving it. Find out which state is experiencing this loss and why.

   ____

   ____

   ____

   ____

   ____

   ____

Gains and Losses in U.S. Population

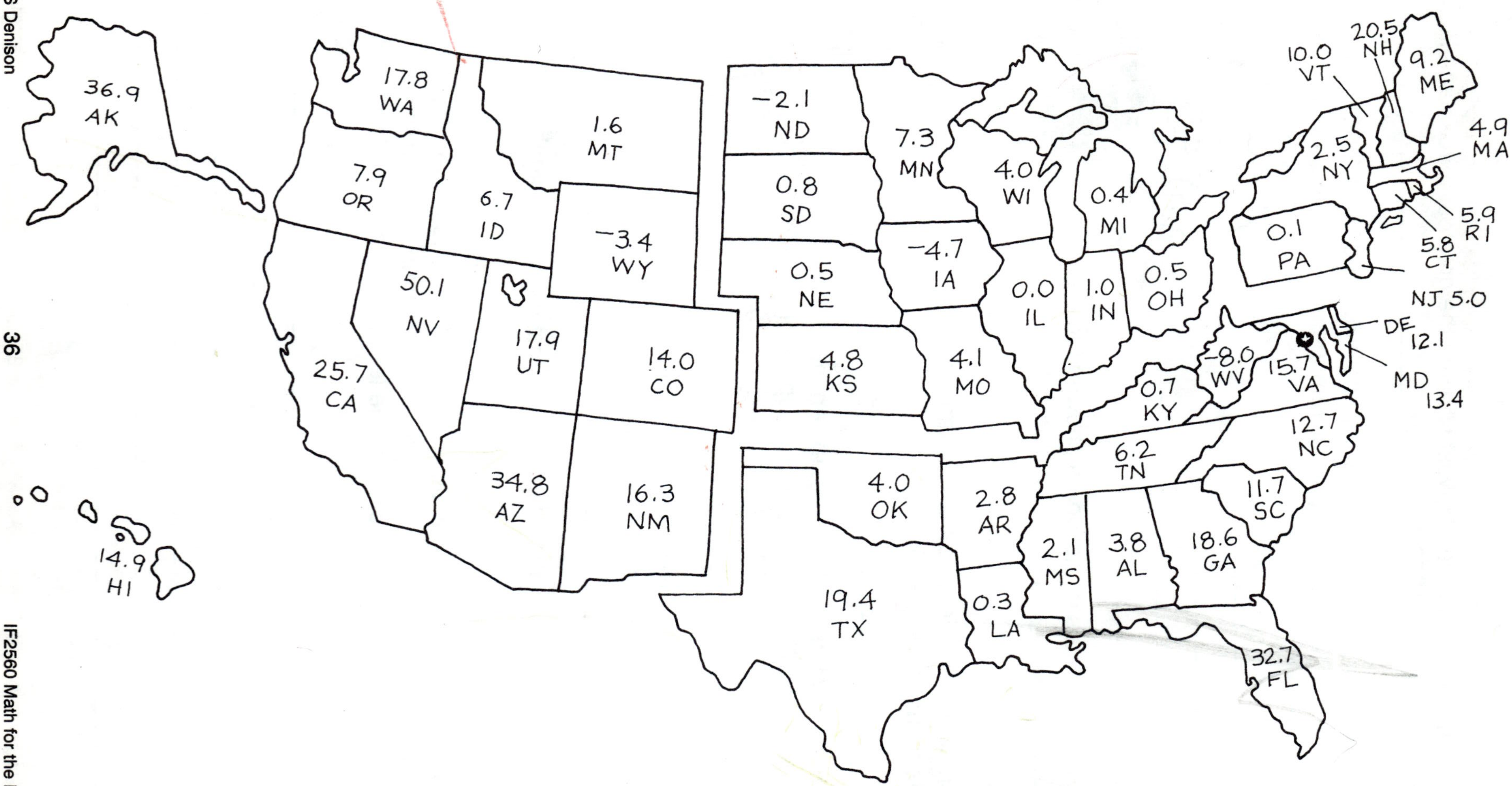

Population gains and losses are expressed in percent. Gains and losses represent 1990 Census compared to 1980 Census.

Name ______________________

The Graying of America

As a nation, the United States grew older during the decade of the '80s. The *median* age increased from 30.0 years to 32.9 years. The *median* is defined as that age when half of the people living in the United States are older and half are younger. This was the largest gain in median age by Americans during this century. What is the significance of this statistic? For one, it means that people are living longer. Analysts also claim that the "Baby Boom" generation moved into middle age during this decade.

The map below shows the median age ranges of all 50 states. Look closely at the map and answer the questions that follow.

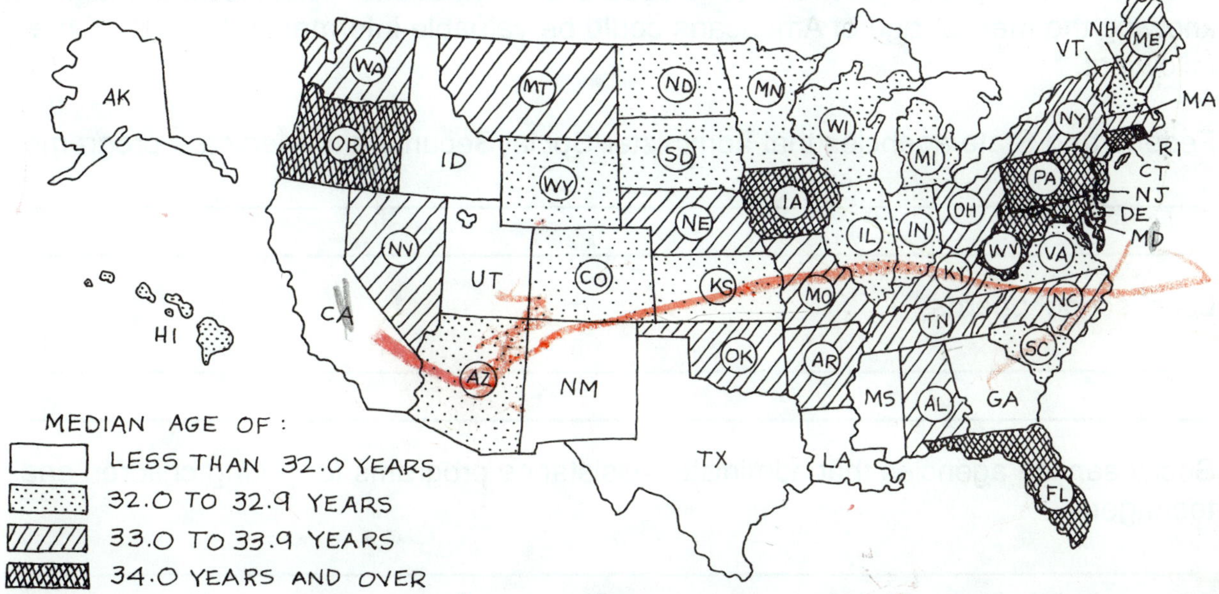

1. Florida had the largest gain in the number of elderly of any state with 18 percent of its population 65 years or older. Can you think of any explanation for this?

____

____

2. Which states have the lowest median age category at less than 32.0 years?

____

3. What general conclusion can you make (if any) about the location where younger and older Americans live?

____

____

____

4. Reference was made in the opening paragraph to the "Baby Boom" generation. Find out the meaning of this term and define it in the space below.

How old would Baby Boomers be today?____________

5. Analysts expect the median age of Americans to be even higher following the Census in the year 2000. They base their predictions on a number of factors. One of those is the Baby Boom generation. Can you predict how this segment of the population will impact the population statistics?

____

____

6. You may consider the question, "Who uses this information?" Think for a moment how knowing the median age of Americans could be valuable information to each of the following:

Federal and State agencies that administer Social Security and Medicare programs—

____

____

Local school boards—

____

____

Social service agencies that administer assistance programs for young children and teenagers—

____

____

Manufacturers of consumer goods—

____

____

7. Who else might consider knowing the median age of Americans "useful information"?

____

____

____

Name ______________________________

Equal Representation

The U.S. Constitution states that the laws of the land shall be made by a legislative branch of government composed of two separate governing bodies. One of these (called the Senate) shall be composed of two persons representing each of these states. The other body (called the House of Representatives) shall have its membership determined by the number of people living in each of the states. The greater the population a state has, the greater the number of members it has in the House of Representatives.

The first House of Representatives in 1789 had 65 members. As the country's population grew, the number of representatives increased. The total membership in the House has been fixed at 435 since the 1910 Census.

Currently, the number of representatives to which each state is entitled is based on the most recent population. The figures represented on the accompanying map are based on the results of the 1990 Census.

Look closely at the information contained on the map and answer the following questions.

1. Which five states have the greatest number of representatives in Congress? List them in descending order (from #1 to #5) and indicate the number of representatives from each state.

#1 ____________________ ______________

#2 ____________________ ______________

#3 ____________________ ______________

#4 ____________________ ______________

#5 ____________________ ______________

2. The Constitution says that each state, regardless of how small it may be, shall have no fewer than one representative in Congress. According to the most recent Census, which states are currently entitled to a single representative?

____

____

3. The Bureau of the Census listed the official population of the United States after the 1990 Census as 248,709,803. If each state truly had "equal representation in Congress," how many people would be represented by each congressman?

____

4. Research the current population statistics for each state that has only a single representative (Question #2). Which of these states has a population fewer in number than the "perfect" size of congressional district you calculated in #3?

____

____

5. Following the 1990 Census, eight states gained additional seats in the House of Representatives. With the size of the House permanently fixed at 435 members, that means some states had to lose seats in Congress. Do a little research on your own and find out who the winners and losers were following the 1990 Census and list them in the space below. (Hint: Find out the number of seats in the House each state had following the 1980 Census; then compare those numbers to the map provided.)

States Gaining House Membership	States Losing House Membership
1 ______	1 ______
2 ______	2 ______
3 ______	3 ______
4 ______	4 ______
5 ______	5 ______
6 ______	6 ______
7 ______	7 ______
8 ______	8 ______
	9 ______
	10 ______
	11 ______
	12 ______
	13 ______

6. How many seats were shifted to different states as the result of the 1990 Census?

____

7. How did your own state fare following the 1990 Census? Did your state gain in Congress, lose, or remain the same? ______________

Gained and Lost Seats in Congress

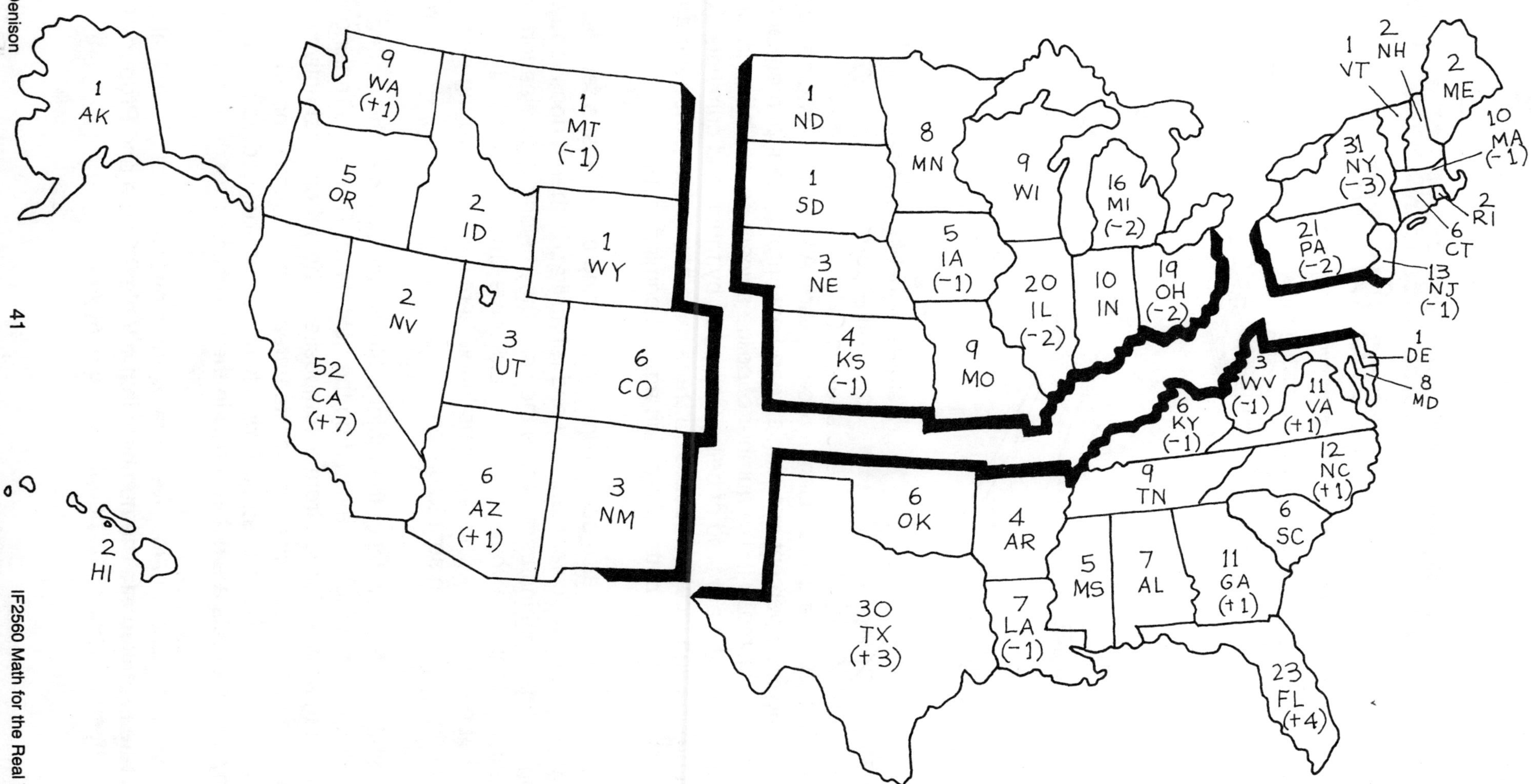

Name ______________________________

Metro Crime

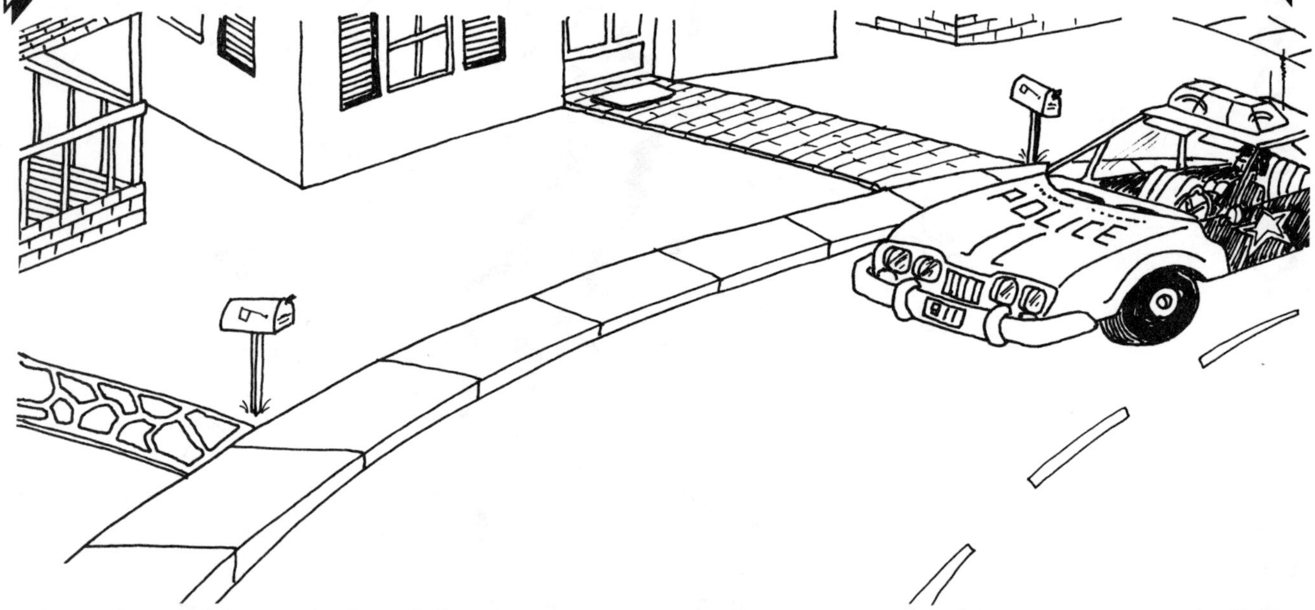

The national crime rate has fallen in recent years. A very encouraging survey reports that overall crime fell 2 percent in a single year! Law enforcement personnel attribute the drop in crime to an increase in the number of police officers as well as a national attitude among people who are tired of being victimized by thugs and criminals. Neighborhood watch groups and local "deputized" citizens have banded together and armed themselves en masse to create a better protection against criminal activity.

The court system has also been helpful in reducing crime. Just a few years ago the system leaned more toward leniency and rehabilitation than it does today. Judges are now putting criminals behind bars, where they say they belong. Age was formerly a factor in determining whether a young person would serve time. Today it is the seriousness of the crime that matters most and age is no longer a major factor. Young people who commit violent crimes are often sent to prison.

Crime statistics for several of the major metropolitan areas of the United States with populations over 100,000 are listed on the next page. The crime ranking of each metropolitan area is derived from a composite of violent criminal activity and property crimes committed within the metropolitan area. *Violent crimes* include murder, rape, robbery, and aggravated assault. *Property crimes* are defined as burglary, auto theft, and larceny. Drug arrests are not included in the statistics.

The lower the number in the overall rank, the higher the rate of criminal activity. The worst crime rate is ranked #1. The numbers in the Violent Crime and Property Crime columns represent crimes per 1,000 people for the past year.

Metropolitan* Area	Rank	Violent Crime	Property Crime
Alabama			
Birmingham	80	11	53
Huntsville	104	7	53
Mobile	68	8	58
Tuscaloosa	12	8	77
Alaska			
Anchorage	35	10	64
Arizona			
Phoenix-Mesa	15	8	75
Tucson	3	8	92
Arkansas			
Fayetteville Springdale	228	2	33
Little Rock	11	14	72
California			
Fresno	10	12	75
Los Angeles-Long Beach	79	15	49
San Francisco	106	9	51
San Diego	117	9	49
Colorado			
Denver	128	7	50
Fort Collins-Loveland	221	3	35
Connecticut			
Hartford	164	5	43
District of Columbia			
Washington, D.C.	138	6	43
Florida			
Broward County	6	10	83
Miami-Dade County	1	19	109
Daytona Beach	112	7	50
Gainesville	2	14	87
Fort Myers-Cape Coral	113	7	51
Palm Beach County	8	11	78
Punta-Gorda	229	3	32
Tallahassee	5	14	83
Georgia			
Atlanta	37	8	65
Savannah	48	5	62
Iowa			
Des Moines	144	3	49
Kentucky			
Louisville	162	7	42
Louisiana			
Baton Rouge	7	13	77
New Orleans	22	12	68
Shreveport	13	10	75
Maine			
Portland	182	3	43
Massachusetts			
Boston	183	7	38
Worcester	210	6	33

Metropolitan Area	Rank	Violent Crime	Property Crime
Michigan			
Detroit	82	10	54
Grand Rapids-Muskegon	158	6	43
Minnesota			
Minneapolis-St. Paul	143	5	47
Mississippi			
Jackson	9	10	78
Nebraska			
Lincoln	52	6	64
Nevada			
Las Vegas	31	12	64
New Hampshire			
Manchester	219	2	36
New Jersey			
Atlantic City	27	7	69
Newark	129	10	46
New York			
New York	70	16	50
Syracuse	227	2	33
North Carolina			
Raleigh-Durham	96	6	56
Ohio			
Cincinnati	171	6	42
Columbus	108	6	53
Oklahoma			
Tulsa	142	8	46
Oregon			
Portland-Vancouver	73	7	58
Pennsylvania			
Johnstown	265	2	15
Philadelphia	187	7	37
Pittsburgh	248	4	26
South Carolina			
Charleston	86	9	54
Myrtle Beach	4	11	87
Tennessee			
Memphis	33	11	64
Nashville	54	11	58
Texas			
Dallas	62	9	59
Houston	123	9	48
San Antonio	34	6	69
Virginia			
Roanoke	220	3	35
Washington			
Seattle-Bellevue	81	5	59
Tacoma	42	9	62
West Virginia			
Wheeling	264	2	20
Wisconsin			
Madison	198	3	39
Milwaukee-Waukesha	141	5	48

*Information for all states was unavailable at time of publication.

1. What two factors are used to measure a city's rank in crime and violence?

____

____

2. Which crimes fall under the category of *violent crimes*?

____

3. Which crimes are classified as *property crimes*?

____

4. Which United States metropolitan area has the dubious "honor" of ranking as the nation's highest in criminal activity per 1,000 persons?

____

5. List the five metro areas in the United States with the highest crime rates.
   1. ____________________
   2. ____________________
   3. ____________________
   4. ____________________
   5. ____________________

6. Do you see any kind of pattern in the location of these five highest ranking metropolitan areas?

____

____

7. Look closely at the list again and list below the areas that ranked #6-#10.

   6. ____________________
   7. ____________________
   8. ____________________
   9. ____________________
   10. ____________________

8. What do all of these metropolitan areas have in common?

____

____

____

9. There are a number of large U.S. cities that have a reputation for being dangerous because of their unusually high rate of criminal activity. Detroit (#83), New York (#70), Washington, D.C., (#138), and Los Angeles (#79) are among those often mentioned on the "dangerous" list. However, the statistics used to compile this report would suggest that these cities might not be as dangerous as earlier indicated. Can you think of any explanation for these cities with past reputations for being dangerous ranking so low on the current "dangerous city" list?

10. Law enforcement officers and public officials from some of the smaller metro areas that rank high on the crime list say that there are some very logical reasons for the numbers . . . and that such statistics are often misleading. Think of some reasons why some of the cities not normally associated with high criminal activity rank so high on the list of this study and discuss them in the space below.

11. Choose any city from the list and find out the population of the entire metro area of that city from the most recent census. Then calculate the number of violent crimes that occurred there last year. Use the same method to determine the number of property crimes. Record your findings and calculations in the space below. Then share your conclusions with other members of your class.

Name ____________________

Education Pays!

We have all been told time after time about the importance of education when it comes to getting a good job someday. The simple truth of the matter is that people who make such comments know exactly what they are talking about. There is a definite correlation between education and wages earned. The more education a person has, the higher the salary that person earns. Below are the plain hard facts that point to this correlation. The data presented was gathered from the most recent census. Look carefully at the data and then answer the questions that follow.

Mean Annual Income of Persons by Educational Attainment, Sex, and Age

Age/Sex	Total # of People	8 yrs or less education	High School	College	College 5 Years or More
Males Total	34,886	19,199	28,043	44,554	55,831
25 to 34	27,743	15,887	24,038	35,534	39,833
35 to 44	37,958	18,379	28,927	47,401	58,542
45 to 54	40,231	19,686	32,862	50,718	62,902
55 to 64	37,469	22,379	30,779	55,518	61,647
65 and +	33,145	17,028	25,516	43,092	52,149
Females Total	22,768	13,322	18,954	28,911	35,827
25 to 34	21,337	11,832	17,076	27,210	32,563
35 to 44	24,453	13,714	19,886	31,631	37,599
45 to 54	23,429	13,490	19,986	29,242	38,307
55 to 64	21,388	13,941	19,382	27,975	33,383
65 and +	19,194	x	18,285	x	x

x = Base figure was too small to meet statistical standards for reliability of a derived figure.

1. At what age group does a man with a high school education make the most money?

____

2. How much money can a male with only an elementary school education expect to make at the age of 40 years? ________________

3. What is the mean income for a male who is between 55 and 64 years old with a college degree? ___________________

4. How much can a female just out of college expect to make during her first few years in the job market? ___________________

5. At what age level does a female with an advanced degree earn her highest salary?

____

6. Compare this figure to the amount a male with an advanced degree makes. At what age level does he command his highest salary? _____________

7. How much more can a 50-year-old male with an advanced degree expect to make than a 50-year-old male with only a high school education?

____

8. How much more does a 44-year-old female make who has a college degree than a 44-year-old female with only a high school diploma?

____

9. How much more does a male college graduate between the ages of 45 and 54 make than a female of the same age range with the same credentials? ________________

10. How much more does a male with an advanced degree between the ages of 55 and 64 years make than a female of the same age range with the same credentials?

____

11. At what age range does it appear that a male reaches his greatest salary potential?

____

12. Does the data in this chart point toward sexual discrimination in the job market when it comes to salary? Explain your answer below by pointing to some of the data presented in the chart to support your answer.

____

____

13. For the sake of comparison, assume that all figures remained the same throughout a worker's career. (We know that in reality this does not happen as inflation, changing annual wages, and a number of other factors enter in.) How much would a male with only an elementary school education typically expect to make between the ages of 25 and 65 (when he could retire)? Use the space below to calculate your answer.

____

14. How much would a male college graduate make during the same time period?

____

15. How much more would the college graduate make during his career than the male with only the eighth grade education? ____________________

16. How much would a female with a high school education make between the ages of 25 and 65 (assuming the numbers in the chart remained the same)?

____

17. How much could a female with an advanced degree make during the same time period (ages 25-65)? ____________________

18. How much more money would the advanced degree mean to the female over her entire career than if she had never gone to college and had only a high school diploma? ____________________

19. Do you begin to see the value of education? More money is important, but what other advantages do those who are better educated have over those who are not?

____

____

Name ______________________________

Heating Our Homes

The most recent census revealed that there were nearly 92 million households in the United States. That figure was up about 12 million over the 1980 figure and represented a nation with nearly 28 million more households than it had following the 1970 Census.

A household, as defined by the Census Bureau, consists of one or more persons who occupy separate living quarters. Separate living quarters are those that have a separate entrance and in which occupants do not live with any other persons in the building. A house, an apartment, or even a single room, if occupied as separate living quarters, is considered a household.

Question H14 on the census form asked people to indicate the fuel used to heat their house or apartment. Below are the data that represent the responses people made to that question.

51% used natural piped gas
26% used electricity
12% used fuel oil
6% used bottled gas
4% used wood, coal, or solar energy
1% used either some other fuel or no fuel at all

1. How many households in the United States are heated by the following fuels?

   natural gas ____________ bottled gas ____________
   electricity ____________ wood, coal, or solar energy ____________
   fuel oil ____________ some other fuel or no fuel ____________

2. What percentage of American households are heated by gas (either piped or bottled)?

   ____

3. Fill in the blanks for this statement: Roughly three out of every four households in American are heated either by ______________________ or ______________________.

4. Charts and graphs provide a much more visual and lasting presentation of certain data than does a mere list. Use the data you were given to portray the same information in a circle or pie graph. Remember that the entire pie represents 100 percent, so the parts marked off must also total 100 percent. Perhaps the best way to proceed is to calculate the portion of the circle that should be marked off and labeled for each of the fuels. Thus 360° x 51% = the size of the circle to be occupied by natural gas.

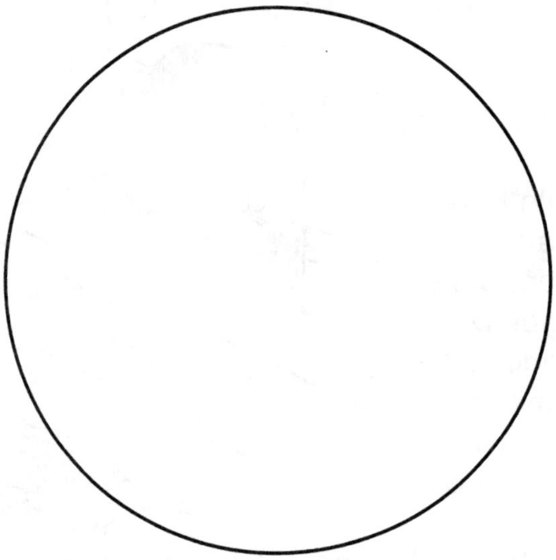

5. Find out how your own home is heated. ______________________ People choose different heat sources for a variety of reasons. Find the reason(s) for the choice made of the heat source that is currently heating your home and discuss it in the space below.

____

____

____

6. Although gas was the leading fuel in the South (45%), almost as many homes (43%) used electricity, the highest percentage of any of the regions. Can you think of any reason for the greater reliance on electricity in the South than in other parts of the country?

____

____

____

Teacher Notes

BUY, SELL, TRADE

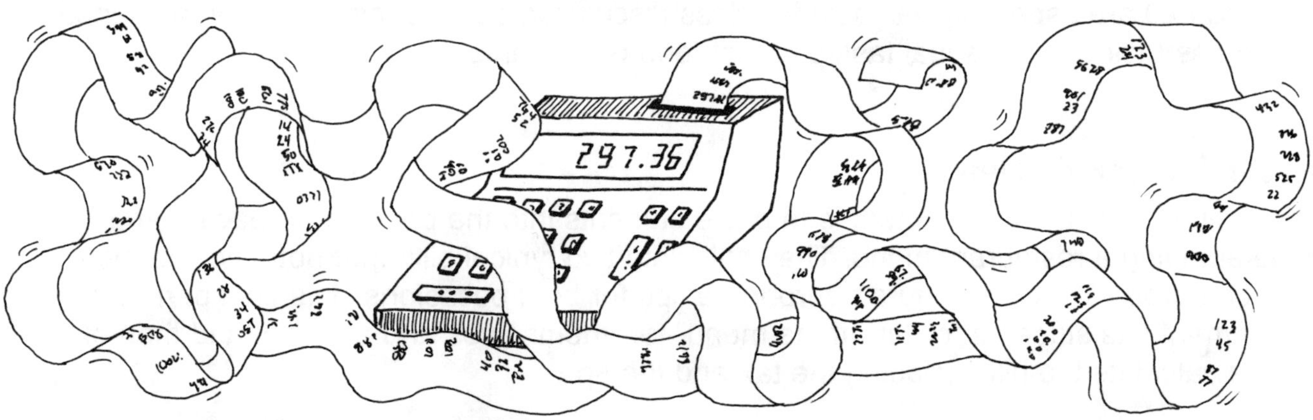

Students at the middle school level are at the age when a lot is starting to happen. Children are no longer completely dependent on their parents. They are beginning to sprout their wings of freedom, and they want to do things on their own and make decisions by themselves.

Helping students to understand some of the good business sense required to function successfully in the real world is really what this entire book is about. In this unit they will learn about comparison shopping and how to measure and calculate the fat and calories in the food prepared at fast-food restaurants. They will "order off the menu" at a restaurant and figure the bill and tip; they will learn about buying a car, buying a home, and buying life insurance. They even get a taste of sports card collecting, though some students may already be involved with this popular hobby.

Throughout the unit, the emphasis is on the math involved. The assignments will require students to perform the math functions that are a part of commonplace reality of the day-to-day events of our world.

Tale of the Tapes

Students are given a newspaper survey of several large competing supermarket prices on a number of food items. They should be placed in pairs and a calculator made available for each pair to complete this activity. Their first task is to find out which store has the least expensive total for all 48 items on the list. They then analyze the list by answering several questions by gleaning information provided by the table. To get them involved in a realistic setting on their own, the last part of the activity has each pair choose their own list, then research two different area stores to find out which offers the better prices. Student interaction is the secret to the success of this assignment, so be certain to allow enough time for each pair to announce the results of their own personal surveys.

Fast Food and Fat

This "healthy" activity focuses on finding and comparing the caloric and fat content of some of the more popular fast-food restaurant chains. Before making the assignment, go over the instruction page with students to make certain they understand what it is they are to do. Take a quick look together at the information provided in the various menus listed in the assignment. Then allow plenty of time for them to make the calculations. Questions #19 and #20 are especially valuable for class discussion as students are given a chance to express their own personal tastes in fast-food restaurants.

Little Anthony's Diner

The main idea of this assignment is to place students into the position of reading and understanding a restaurant menu. They calculate two typical bills that both include more than one order. They also figure appropriate tips for both occasions. The final part of the assignment lets students "order off the menu" for themselves and a friend, and then of course calculate the bill, including the tax and the tip.

Tyler's Sports Cards

There may already be students in your class enjoying the hobby of collecting sports cards (or perhaps some other form of trading cards). If so, invite those students to share with others their current collections. The activity presents a simulation of a young boy who has turned his desire for sports cards into more than a casual hobby. The problems involve figuring percentages, calculating the odds of getting specific insert cards (cards of greater value), and determining current values of sports cards.

Laurie's Laredo

While middle school students do not buy cars, many are already interested in them and are very much concerned about how much they cost. Students experience Laurie's problems and joys of buying a car. Interpreting a new car sticker price, investigating "buying by invoice," and finding ways to sell her older car are but a few of the realities associated with buying a new car.

House for Sale

Students investigate home buying in this activity that gets them involved in calculating and comparing various rates for home mortgage loans. The importance of "shopping around" for the best rate available points to the savings that can be realized by just a fraction of a percentage point. Students also see how much money

can be saved by paying off a home mortgage in 15 years rather than 30 years. A great addition to this lesson would be to have a real estate agent speak to the class about home buying.

A Fact of Life

The final activity in this unit brings focus to buying term life insurance. While buying life insurance is not exactly a focal point in the minds of middle school students, they will learn about its importance as they grow older. They learn how to calculate rates based on a table provided for various ages. The table also points to the significant difference in rates between smokers and nonsmokers. Students will also see how the rates escalate dramatically for older people.

Name ____________________

Tale of the Tapes

Newspapers in several American cities conduct surveys from time to time comparing prices of competing stores. Such surveys are popular among consumers because they give them a feel for which stores are more expensive and which stores offer greater values. The results of a recent survey of eight supermarkets appear below. There were 48 items on the "shopping list." Those products chosen were brand name items that would be found in every store. Look closely at the list of prices; then answer the questions that follow on the next page.

	Super K-mart	Albertson's	Safeway	ABCO	Smith's	Smitty's	Fry's	Bashas'
Grade AA Eggs large, 1 dozen, cheapest	$1.17	$1.13	.89	$1.10	$1.11	$1.09	$1.09	.79
1 gallon skim milk, cheapest	2.09	1.99	1.95	1.89	1.89	1.89	1.89	1.89
Hunt's Pudding Cups, 4 pack	1.29	1.39	1.29	1.39	1.39	1.25	1.49	1.49
Kraft American Singles, 24 slices	3.39	3.89	3.79	3.79	3.99	3.79	3.79	3.79
Parkay Margarine, 16 oz. sticks	.65	.85	.69	.89	.99	.85	.89	.85
Lipton Rice & Sauce, Cheddar-Broccoli 4.5 oz.	1.17	1.09	.99	1.29	1.29	1.29	1.19	1.25
Gold Medal All-Purpose Four, 5 lb.	1.55	1.55	1.29	1.69	1.89	1.69	1.69	1.59
C&H Pure Cane Sugar, 5 lb.	2.29	1.99	1.99	2.29	2.29	2.29	2.19	2.29
Brawny Thirst Pocket Paper Towels, Single Roll	1.09	.99	1.09	1.09	.99	.79	1.09	1.09
Charmin Toilet Tissue, 4 pack	1.49	1.39	1.49	1.49	1.29	1.49	1.49	1.49
Grant's Farm Stone Ground Wheat Bread, 24 oz.	1.85	1.89	1.99	1.99	1.99	1.99	1.99	1.99
Golden Delicious Apples, 3 lbs.	2.67	2.67	2.37	3.57	2.37	2.97	2.37	1.47
Carrots, 1lb. bagged or loose	.45	.45	.44	.44	.50	.33	.50	.45
Creamette Spaghetti, 16 oz.	.79	.75	.99	.59	.69	.79	1.19	.89
Prego Traditional Spaghetti Sauce 28 oz.	2.25	2.19	1.99	2.29	1.89	1.50	2.29	2.29
Campbell's Chicken Noodle Soup 10¾ oz.	.53	.49	.53	.53	.55	.53	.53	.53
Del Monte Canned Cut Green Beans, 14½ oz.	.63	.69	.69	.50	.65	.69	.69	.73
Sunny Delight, 9 pack	2.89	2.69	2.89	1.99	2.89	2.89	2.89	3.29
Ocean Spray Cran-Apple Juice, 64 oz. bottle	3.59	3.59	3.59	3.49	3.55	3.59	3.59	3.55
Downflake Waffles, 16 economy pack	2.03	1.99	2.49	2.19	2.29	2.49	2.49	2.49
Log Cabin Lite Syrup, 24 oz.	3.39	3.39	3.09	3.79	3.49	3.49	3.49	3.69
Kellogg's Pop Tarts Frosted Strawberry, Box of 10	2.09	1.89	1.99	1.99	1.99	2.09	2.09	2.09
Cheerios, 20 oz.	4.48	4.29	4.49	4.99	4.77	4.59	4.49	4.55
Chips Ahoy Cookies, 18 oz.	2.69	2.50	3.19	2.89	3.09	2.50	2.99	2.99
Puffs Plus Facial Tissue, 70 count	1.39	1.29	1.39	1.39	1.49	1.39	.99	1.29
Maxwell House Coffee, 13 oz.	3.29	3.09	2.49	3.49	3.19	3.29	3.49	3.59
Coffeemate Creamer Vanilla, 7 oz. powder	1.29	1.89	2.09	1.99	2.13	2.19	2.09	2.09
Tostitos Baked Chips, 8 oz.	1.50	1.73	1.99	.99	1.65	1.99	1.99	.99
Pace Picante Sauce, 16 oz.	2.09	1.99	2.09	2.19	2.19	2.09	2.09	2.09
Oscar Mayer Beef Bologna, 12 oz.	2.59	2.49	2.49	2.59	2.69	2.49	2.49	2.49
Coca-Cola Cans, 12 pack	3.59	3.79	3.49	3.59	2.99	3.59	2.79	4.58
Russett Potatoes, 10 lb. bag	1.99	2.29	1.99	2.49	1.29	2.49	2.29	2.49
Miracle Whip, 32 oz.	1.67	1.99	1.99	2.39	2.09	2.19	2.19	1.99
Vlasic Kosher Dill Spears, 24 oz.	2.39	2.49	2.59	2.79	2.79	2.59	2.59	2.59
A-1 Steak Sauce, 10 oz.	3.56	3.69	3.99	3.99	4.23	4.49	4.39	4.19
Jimmy Dean Pork Sausage, Reg. 12 oz	2.89	2.39	2.59	2.29	2.59	2.59	2.59	2.79
T-Bone Steak, per lb.	3.17	4.89	4.99	3.99	5.99	4.89	4.89	4.89
Whole young fryer chicken, per lb.	.83	.79	.89	.79	.89	.89	.89	.99
Downy Ultra Fabric Softener, 40 oz.	4.79	4.79	4.59	5.29	4.99	5.09	4.99	4.99
Fresh Scoop Unscented Cat Litter, 7.5 lb	3.57	3.99	3.99	2.69	3.99	3.99	3.69	3.99
Kibbles & Bits Dog Food, 4 lb. bag	4.19	3.99	4.19	4.39	4.45	4.29	4.19	4.19
Red Baron Pepperoni Deep Dish Singles, 2 pack	2.89	2.69	2.99	2.79	2.99	2.99	2.99	2.99
Breyers Vanilla Ice Cream, ½ gallon	4.59	4.49	4.49	4.69	3.59	4.33	4.49	4.49
Hershey's Chocolate Syrup, 24 oz	1.86	1.65	1.79	1.79	1.85	1.79	1.79	1.79
Crisco Vegetable Oil, 48 oz.	2.89	2.89	2.99	2.19	2.45	2.99	3.19	3.19
Knudsen Lowfat Cottage Cheese, 16 oz.	1.46	1.69	1.69	1.69	1.89	1.69	1.69	1.69
Gerber Rice Cereal For Babies, 8 oz. box	1.19	1.09	.99	1.25	1.27	1.19	1.19	1.19
Sunlight Liquid Dish Soap, 22 oz.	1.39	1.19	.99	1.39	1.39	1.39	.99	1.29

1. Your task is to use a calculator to add up all totals. You should use a calculator that has a printout tape if at all possible. Then you will have a check of the figures on the tape with those in the survey summary. Place your total cost of the items for each store in the blank spaces provided below:

Super K-mart ____________

Albertson's ____________

Safeway ____________

ABCO ____________

Smith's ____________

Smitty's ____________

Fry's ____________

Bashas' ____________

2. Which store won this "tale of the tapes"? ________________

3. Which store had the highest total price for the 48 items used in this survey?

____

4. How much difference was there between the lowest priced store and the store that finished in second place in the survey? ________________

5. How much money could be saved by buying the 48 survey items in the lowest priced store rather than buying them in the store with the highest total? ____________

6. Which two stores were the closest competitors in the total cost of the items listed in the survey? ________________________________

7. Which store had the lowest price on T-bone steak?____________ Which store had the highest price? ______________ How much difference was there per pound between the highest price and the lowest price? ____________

Can you think of any possible reason for there being such a difference in price between the two stores? ________________________________

____

8. Which stores had eggs on sale? ____________________

9. Which store had the best price on apples? ____________________

10. Which store had the best price on Coca Cola? ____________________

11. Which store had the lowest price on a gallon of skim milk of any brand? ________

____

12. How much difference was there in price between a can of Campbell's chicken noodle soup at Albertson's and Smith's? ____________________

13. How much could be saved by buying a jar of Prego spaghetti sauce at Smitty's rather than buying the same jar at ABCO? ____________________

14. Which stores have Vlasic kosher dill spears for $2.59? ____________________

____

15. Why is Fry's sale price on Fresh Scoop unscented cat litter not such a "sale" after all?

____

16. If chips and picante sauce were your entire shopping list, where should you go?

____

17. How much difference in price was there on Kellogg's PopTarts between Smitty's and Bashas'? ____________________

18. Which store had the best price on fabric softener? ____________________

19. Which stores had the best buy on sugar? ____________________

____

20. Which stores had the highest price on cran-apple juice?____________________

____

21. In the space below choose any ten items that appeal to your good taste. Simply list the items; then compare what they would have cost at each of the stores and find out which store would have won your own "tale of the tape." Share your choices and results with other members of your class.

22. Can you think of any reasons (other than lower prices) for people choosing to shop at specific supermarkets? Think of reasons why your own family buys more at one store than another. Share your thoughts with other members of your class.

____

23. For this last part of the activity, you need to have a partner who can help you do some of the legwork. Your task is to choose 15 items from the list used in the survey. You can make substitutions of your own if you prefer. Jot down your list below. Then you and your partner each need to go to a separate store and jot down how much each of the items on your list would cost at that particular store. Then get together with your partner and compare prices. Which store won the "tale of the tape"? Share the results of your survey with other partner teams in your class.

	ITEM	STORE ________	STORE ________
1.			
2.			
3.			
4.			
5.			
6.			
7.			
8.			
9.			
10.			
11.			
12.			
13.			
14.			
15.			

Fast Food and Fat

It seems that today's fast-moving lifestyle keeps us going at such a "full speed ahead" clip that we barely have enough time for eating and sleeping! This breakneck pace has brought about the immense popularity of fast-food restaurants all over America. McDonald's, of course, was among the first, and remains the largest and most successful. But other major fast-food chains have also become very popular among the eat-on-the-go crowd. Burger King, Kentucky Fried Chicken (KFC), Pizza Hut, Hardee's, Arby's, Domino's Pizza, Boston Market, Little Caesar's, Subway, Wendy's, and Taco Bell have all firmly established themselves in this lucrative food industry.

But there is a downside to all this convenience food. Much of it is loaded with calories and fat! Americans have become consciously aware of the dangers of eating too much and of eating foods filled with fat. This group does not just include middle-aged and older people either! Parents who have weaned themselves from fatty food diets have also encouraged their children to "eat healthy."

Because of this, fast-food franchises have had to alter their menus to offer additional choices that have less fat and fewer calories. You will still find lots of high fat foods at these places, but there are now healthy alternatives available for those who want them. After all, these food-on-the-go restaurants do not want to lose any business!

Here's how it all works. If you are watching your weight and keeping your calories to a minimum of 1,200 per day, you need to count the number of calories you consume each time you ingest food or drink into your system. You need to remember as you progress through the day just how many calories you have consumed, so you will know how many you have remaining. But you also need to know the amount of fat in the foods you eat. Nutrition experts suggest that you should limit your fat calories to roughly 30 percent of your total daily calorie count. Thus a 1,200 calorie diet should contain only 360 fat calories (1,200 x 30%).

1. If one "fat gram" contains 9 calories, then what would be the approximate number of fat grams in the above diet for an entire day? __________

2. If you then divide that number of fat grams equally among three meals, each meal should consist of food that totals only ________ grams of fat. Obviously, all meals will not work out exactly to the number of allowed calories or grams of fat. But keeping a running total of both will help the person on the diet to make choices that will keep the day's end total consistent with the suggested allowances.

The list below and on the next page will give you menu items from some of the more popular fast-food chains and some of the options people have when they go to such restaurants. Some of the choices are healthy; some are not. The questions that follow the list can be answered by analyzing the data given.

Arby's	CALORIES	FAT(G)
Main Dishes		
Best Bets		
Light Menu—Roast Beef Deluxe	296	10
Light Menu—Roast Chicken Deluxe	276	6
Light Menu—Roast Turkey Deluxe	260	7
Roast Chicken Salad	149	2
Timberline Chili	220	10
Worst Bets		
Bacon 'n Cheddar Deluxe	539	34
Chicken Cordon Bleu	623	33
Giant Roast Beef Sandwich	555	28
Italian Sub	675	33
Philly Beef 'n Swiss	755	47
Roast Beef Sub	700	42
Triple Cheese Melt	720	45
Roast Chicken Club	546	31
Side Dishes		
Best Bets		
Garden Salad	61	1
Lumberjack Mixed Vegetable Soup	90	4
Old-Fashioned Chicken Noodle Soup	80	2
Plain Baked Potato	355	1
Side Salad (no dressing)	23	1
Worst Bets		
Baked Potato w/ Margarine & Sour Cream	578	24
Broccoli 'n Cheddar Baked Potato	571	20
Cheddar Curly Fries	333	18
Deluxe Baked Potato	736	36

Boston Market	CALORIES	FAT(G)
Main Dishes		
Best Bets		
¼ Chicken, Dark Meat (no skin)	210	10
¼ Chicken, White Meat (no skin or wing)	160	4
Chicken Sandwich (no cheese or sauce)	430	4
Ham & Turkey Club Sandwich (no cheese or sauce)	430	6
Ham Sandwich (no cheese or sauce)	450	9
Skinless Rotisserie Turkey Breast	170	1
Turkey Sandwich (no cheese or sauce)	400	4
Worst Bets		
Caesar Salad Entree	520	43
½ Chicken (with skin)	630	37
Chicken Caesar Salad	670	47
Ham & Turkey Club Sandwich w/Cheese & Sauce	890	43
Ham Sandwich w/Cheese & Sauce	760	36
Side Dishes		
Best Bets		
Fruit Salad	70	*1
Steamed Vegetables	36	*1
Worst Bets		
Tortellini Salad	380	24

*less than 1 gram

Burger King	CALORIES	FAT(G)
Main Dishes		
Best Bets		
Broiled Chicken Salad	200	10
Chicken Tenders	250	12
Worst Bets		
BK Big Fish Sandwich	720	43
Chicken Sandwich	700	43
Croissan'wich w/Bacon, Egg, and Cheese	350	24
Croissan'wich w/Ham, Egg, and Cheese	350	22
Croissan'wich w/Sausage, Egg, and Cheese	530	41
Double Cheeseburger	600	36
Double Cheeseburger w/Bacon	640	39
Double Whopper Sandwich	870	56
Double Whopper w/Cheese Sandwich	960	63
French Toast Sticks	500	27
Whopper Sandwich	640	39
Whopper w/Cheese Sandwich	730	46
Whopper Jr. w/Cheese Sandwich	460	28
Side Dishes		
Best Bets		
Garden Salad (no dressing)	90	5
Side Salad (no dressing)	50	3
Worst Bets		
French Fries (medium)	400	20
Hash Browns	220	12
Onion Rings	310	14
Condiments		
Best Bets		
Reduced Calorie Light Italian Dressing	15	1
Worst Bets		
Bleu Cheese Dressing	160	16
Ranch Dressing	180	19
Thousand Island Dressing	140	12

Domino's Pizza	CALORIES	FAT(G)
Main Dishes		
Best Bets		
Cheese 12" Hand-Tossed Pizza (2 slices)	345	10
Ham 12" Hand-Tossed Pizza (2 slices)	362	10
Veggie 12" Hand-Tossed Pizza (2 slices)	360	10
Worst Bets		
Italian Sausage & Mushroom Deep Dish Pizza (2 slices)	618	28
Pepperoni Deep Dish Pizza (2 slices)	622	29
X-tra Cheese & Pepperoni Deep Dish Pizza (2 slices)	671	33

Kentucky Fried Chicken (KFC)	CALORIES	FAT(G)
Main Dishes		
Best Bets		
Original Recipe Chicken (1 drumstick)	130	7
Rotisserie Gold Chicken (breast and wing, skin removed)	199	6
Worst Bets		
Colonel's Chicken Sandwich	482	27
Extra Tasty Crispy Chicken Breast	470	28
Extra Tasty Crispy Chicken Thigh	370	25
Hot & Spicy Chicken Breast	530	35
Hot & Spicy Chicken Thigh	370	27
Hot Wings Pieces	471	33
Side Dishes		
Best Bets		
BBQ Baked Beans	132	2
Garden Rice	75	1
Green Beans	36	1
Mean Greens	52	2
Red Beans and Rice	114	3
Worst Bets		
Biscuit	200	12
Corn on the Cob	222	12

Little Caesars	CALORIES	FAT(G)
Main Dishes		
Best Bets		
Caesar Salad	140	5
Pan!Pan! Cheese Only (1 slice, med.)	181	6
Pan!Pan! Pepperoni (1 slice, med.)	199	8
Pizza!Pizza! Cheese Only (1 slice, med.)	201	7
Pizza!Pizza! Pepperoni (1 slice, med.)	220	9
Worst Bets		
Cold Deli-Style Ham & Cheese Sandwich	728	35
Cold Deli-Style Italian Sandwich	740	37
Cold Deli-Style Veggie Sandwich	647	29
Hot Oven-Baked Cheeser Sandwich	822	39
Hot Oven-Baked Measta Sandwich	1,036	56
Hot Oven-Baked Pepperoni Sandwich	699	47
Hot Oven-Baked Supreme Sandwich	894	46
Side Dishes		
Best Bets		
Crazy Bread (1 piece)	106	3
Individual Tossed Salad (no dressing)	116	3
Worst Bets		
Antipasto	176	12

McDonald's	CALORIES	FAT(G)
Main Dishes		
Best Bets		
Chef Salad	210	11
Chunky Chicken Salad	160	5
Hamburger	270	9
Hotcakes (plain, syrup, no margarine)	280	4
McGrilled Chicken Classic	250	3
McLean Deluxe	340	12
Scrambled Eggs	170	12
Worst Bets		
Bacon, Egg & Cheese Biscuit	450	27
McChicken Sandwich	490	29
Quarter Pounder w/Cheese	520	29
Sausage Biscuit	430	29
Sausage Biscuit w/Egg	520	35
Sausage McMuffin w/Egg	440	29
Side Dishes/Desserts		
Best Bets		
Apple Bran Muffin	180	1
English Muffin	140	2
Side Salad (no dressing)	45	2
Strawberry Lowfat Frozen Yogurt Sundae	240	1
Vanilla Lowfat Frozen Yogurt Cone	120	1
Worst Bets		
Apple Danish	360	16
Biscuit	260	13
Cheese Danish	410	22
Cinnamon Raisin Danish`	430	22
French Fries ("super size")	540	26
Raspberry Danish	400	16
Condiments		
Best Bets		
Barbecue Sauce	50	0
Honey	45	0
Sweet 'n Sour Sauce	50	0
Worst Bets		
Bleu Cheese Dressing (1 pkg)	190	17
Ranch Dressing (1 pkg)	230	21

Pizza Hut	CALORIES	FAT(G)
Main Dishes		
Best Bets		
Cheese Bigfoot Pizza (1 slice)	186	6
Ham Hand Tossed Pizza (1 slice)	213	5
Pepperoni Personal Pan Pizza (whole pie)	637	28
Pepperoni Bigfoot Pizza (1 slice)	205	7
Super Supreme Pan Pizza (1 slice)	323	17
Veggie Lover's Thin 'n Crispy Pizza (1 slice)	186	7
Worst Bets		
Meat Lover's Pan Pizza (1 slice)	340	18

Taco Bell	CALORIES	FAT(G)
Main Dishes		
Best Bets		
Light Bean Burrito	330	6
Light Burrito Supreme	350	8
Light 7-Layer Burrito	440	9
Light Soft Taco	180	5
Light Soft Taco Supreme	200	5
Light Taco	140	5
Light Taco Salad w/o Chips	330	9
Light Taco Supreme	160	5
Soft Taco	220	11
Taco	180	11
Worst Bets		
Burrito Supreme	440	19
Light Taco Salad w/Chips	680	25
7-Layer Burrito	540	23
Taco Salad	860	55
Side Dishes		
Best Bets		
Seasoned Rice	110	3
Worst Bets		
Nacho Chips	345	18

Subway	CALORIES	FAT(G)
Main Dishes		
Best Bets		
6" Sandwiches (on wheat bread):		
Club	330	6
Ham	273	4
Roast Beef	299	6
Steak & Cheese	363	10
Subway Melt	348	11
Turkey	276	4
Turkey & Ham	275	4
Veggies	223	3
4" Round Sandwiches:		
Ham	209	3
Light Seafood & Crab	230	5
Roast Beef	233	4
Turkey	219	3
Salads		
Club Salad	123	3
Light Seafood & Crab Salad	155	8
Turkey Salad	99	2
Veggies Salad	45	1
Worst Bets		
6" Sandwiches:		
Classic Italian BMT	574	35
Meatball	411	15
Seafood & Crab	415	19

3. Looking at the menu of several of the possibilities at Arby's, which appears to be the most healthful choice from the "Main Dishes" part of the menu? ____________ ____________ How many calories?____________ How many grams of fat? _______

4. How much worse (in calories and grams of fat) a "food sin" is a Deluxe Baked Potato than a Plain Baked Potato? Calories__________ Grams of Fat_______

5. What choices would you have from the menu at Boston Market if you were eating lunch there and wanted to stay within the 1,200 calories/day and 40 grams of fat target plan? ____

____

____

6. If you go really "big time" at Burger King and treat yourself to a Double Whopper with Cheese Sandwich, what do you have left in the way of calories and grams of fat for the remainder of the day?____

____

7. We all think of ordering salads as good choices "for health's sake." What is wrong with ordering a Garden Salad with Ranch Dressing at Burger King? ____________

____

8. Which is better for your health at KFC, a Hot & Spicy Chicken Breast with Red Beans and Rice or a Colonel's Chicken Sandwich and Baked Beans? ____________

____

9. Why do you suppose Italian Sausage & Mushroom Deep Dish Pizza is such a bad choice at Domino's? ____

____

10. The worst bet on the menu at Taco Bell is a Taco Salad. Have you ever eaten one? If not, investigate the reason it is such a bad choice, both in terms of high calories and high fat content, and jot down your answer.____

____

____

11. When it comes to Subway, which of the 6" Sandwiches is the best choice, both in terms of low calorie content and low fat content? ____________

12. We often attach a healthful connotation to eating seafood. Why do you suppose the only two seafood sandwiches at Subway are among the worst bets there?

____

____

13. Suppose you stop by Pizza Hut for one of those quick "30-minute" lunches and order a Pepperoni Personal Pan Pizza. How many calories and grams of fat do you have left in your day to remain within the suggested dietary standard discussed earlier?

____

14. What are three good choices at McDonald's if you are eating there for lunch?

____

____

15. If you and your family ate dinner at Pizza Hut and you ordered a Super Supreme Pan Pizza and ate two slices, how many calories and grams of fat would you consume?

____

16. What is the difference in both calories and grams of fat between a medium-size order of French Fries at Burger King and an order of super size French Fries at McDonald's?

____

17. Name three good choices as side dishes at KFC.

____

18. At Taco Bell, a Light 7-Layer Burrito contains 1,430 milligrams of sodium (salt), which translates to 60 percent of the daily value for sodium. Based on this information, what is 100 percent of the daily value of sodium?

____

This explains why salt should also be considered when developing a diet plan for fast-food restaurants.

19. Just looking at the portions of the menus found on these pages, which of these fast-food restaurant chains seems to have the best offering for people who are counting their calories and fat grams? _______________________

____

20. In the space that remains, jot down the foods you selected the last time you ate at one of these major fast-food restaurants. Even though you were not counting calories at the time, provide both a calorie count and fat grams consumed count for the food you consumed.

____

____

Name ______________________________

Little Anthony's Diner

Little Anthony's Diner is definitely the place to be! It has great sandwiches and soft drinks (It even serves Green Rivers!) and the prices are absolutely affordable. But the biggest attraction of Little Anthony's is the place itself. The theme is the 1960s when rock-'n-roll was alive and well. Today, many of those tunes that were so popular during the '60s have reemerged and are popular once again in the '90s!

Some are recorded by current rock groups doing their own renditions, but many are the sounds of the original artists of that day who live on in the juke boxes, on the rock music stations, and in the hearts of millions of Americans. The sound system at Little Anthony's belts out tunes of the day from 11:00 a.m., when it opens, until 12:00 midnight, when it closes. There's a disk jockey on duty at all times and the requests are free.

The walls are covered with posters and other memorabilia of the '60s. Many of the booths are "re-made" versions of automobiles that were popular during the '60s (some with hoods and dashboards!). The waitresses and waiters roller skate up to the table to take your order, and the black and white checkerboard tile dance floor welcomes "twisters and shouters" any time! There's a twist contest every Wednesday and Saturday night, and hula hoop contests are held every Friday. Many patrons look forward to Limbo Tuesday, when they can show "how low they can go!"

T-shirts are only $10.00 and giant Coke mugs with Little Anthony logos sell for just $4.00. The menu on the next page will show you just a few of the reasons for Little Anthony's appeal. Look closely at the menu and answer the questions that follow.

Little Anthony's Diner

THE LITTLE BE-BOPPER

CHICKEN NUGGETS & CHIPS OR FRIES $2.95
BURGER & CHIPS OR FRIES $2.95

ICE CREAM

Sundaes $2.50
Hot Fudge, Strawberry, Butterscotch, Pineapple
Hot Fudge Banana Extravaganza $3.25

SODA POP

"Bottomless Glass" $1.25
Pitcher $4.25
Coke - Sprite - Ginger Ale - Diet Coke - Root Beer - Green River

EATIE GOURMET BURGERS 6 oz. of Fresh 100% Choice Angus Beef, Charcoal Broiled, Served with Fries, Lettuce, Onion, Tomato, and Pickle

JAIL HOUSE ROCK $4.50
Plain Burger
CHEEZIE BURGER $4.75
Burger with American, Swiss, Jack, or Cheddar
PATTY MELT $4.95
Burger with Grilled Onions, American Cheese on Rye
BE BOP BACON BURGER $4.95
Burger with Bacon, American, Swiss, Jack, or Cheddar
C.C. RYDER $5.20
Burger with Chili, Cheese, and Onions
DAD'S FAVORITE $4/05
Burger with Grilled Onions
WILD THING $4.95
Burger with Sauteed Mushrooms (with Cheese, Add .25)
CALIFORNIA DREAMIN' $5.20
Burger with Sauteed Mushrooms, Swiss, and Avocado
LA BAMBA BURGER $4.95
Burger with Green Chile Strips, Jack Cheese, and Salsa

CHICKEN DELIGHTS

GRILLED CHICKEN SANDWICH $4.95
Grilled Chicken Breast, Choice of Bread
CALIFORNIA DREAMIN' CHICKEN SANDWICH $5.20
Grilled Chicken Breast with Sauteed Mushrooms, Swiss Cheese, and Avocado
BE BOP BACON CHICKEN SANDWICH $4.95
Grilled Chicken Breast with Bacon, American Swiss, Jack, or Cheddar

THINGS ELVIS WOULDN'T EAT!

THE LITE ONE $3.95
1/2 Sandwich with a Cup of Soup, Dinner Salad, or Low-Fat Cottage Cheese
TUNA SALAD $5.25
On a Bed of Lettuce with Tomato Wedges and Cucumber Slices w/Dinner Roll
Ask about our non-fat dressings!
CHARBROILED CHICKEN SALAD $5.25
Tender Chicken Breast Served on Top of a Crisp Garden Salad w/Dinner Roll. Loaded!
Ask about our non-fat dressings!
GRILLED CHICKEN PLATTER $3.95
Served with Steamed Vegetable Medley and a Dinner Roll.
SOUP AND SALAD $3.50
Bowl of Soup and Dinner Salad with Choice of Dressing and Crackers or Roll.
GARDEN BURGER $4.50
Meatless Combination of Vegetables and Grains and Our Homemade Bun Served with Fries. Cheese, Add .25.

WILD THING CHICKEN SANDWICH $4.95
Grilled Chicken Breast with Sauteed Mushrooms (with Cheese, Add .25)
BBQ CHICKEN SANDWICH $4.95
Chicken Breast in a Tangy Sweet BBQ Sauce Served Open Face w/Fries
LA BAMBA CHICKEN $4.95
Grilled Chicken Breast with Green Chile Strips, Jack Cheese, and Salsa
TERIYAKI CHICKEN $4.95
Grilled Chicken Breast in Our Special Teriyaki Sauce and a Pineapple Slice!

SANDWICHES ETC.

BIG CHEESE .... #3.25
Grilled Cheese (with Tomato, Add .25)(with Bacon, add .50)
BBQ BEEF SANDWICH $4.50
Choice Top Round in a Tangy Sweet BBQ Sauce Served Open Face! w/Fries
PIG IN A GARDEN .... #3.95
Bacon, Lettuce, and Tomato with Mayo
HOME ON THE RANGE $3.95
Roast Beef (We Cook It Here.)
RUN AROUND SUE $3.95
Turkey (Roasted Turkey Breast)
TUNA SALAD SANDWICH $3.95
Fresh Homemade Tuna Salad Sandwich w/Fries—Just Like Mom's
EGG SALAD SANDWICH $3.95
Fresh Homemade Egg Salad Sandwich w/Fries—Mom's Again!
FISHERMAN $5.50
Icelandic Cod Deep Fried in a Beer Batter w/Fries and Tartar Sauce
CHUBBY CHECKER TRIPLE DECKER $4.95
The All-American Club
STEAK SANDWICH $5.95
Served on a French Roll with Sauteed Mushrooms with Fries and Garnished with Onion Rings!
CHICKEN STRIPS $4.25
Deep Fried Chicken Breast Strips with Fries and BBQ Sauce
FRENCHIES' FAVORITE $5.25
French Dip (Freshly Roasted Top Round with Swiss Cheese and Sauteed Mushrooms and Au Jus)
SURFER'S DELIGHT $5.50
Shrimp Basket (Delicious Deep Fried Shrimp) w/Fries
HOPPIN' HAM SANDWICH $4.25
Baked Ham on Choice of Bread w/Fries
LITTLE ANTHONY'S MELT $4.95
Grilled Ham, Turkey, or Tuna with Your Choice of Bread and Cheese. Tomato, Add .25. Bacon, Add .50. Avocado, Add .75.

DOGS

HOUND DOG $3.95
1/4 lb. Hebrew National Hot Dog with Relish and Fries
WORKIN' DOG $4.25
1/4 lb. Hebrew National with Tomato, Cheese, Onion, and Fries
BELLEY BURNER $4.50
1/4 lb. Hebrew National with Chili, Cheese, Onion, and Fries

GRANDMA TONY'S PIZZA

Sausage, Ham, Pepperoni, Jalapenos, Green Peppers, Onions, Mushrooms, Diced Tomatoes, Pineapple, Black Olives $9.00 Medium $12.50 Large

1. Rhonda and her boyfriend Tim and Maria and her boyfriend Raol are headed for a Friday date at Little Anthony's Diner. When they order food, they tell the waiter to bring two separate bills—one for Rhonda and Tim and another for Maria and Raol. The following orders are taken: Rhonda—Wild Thing sandwich and a Ginger Ale, Tim—Chubby Checker Triple Decker and a Green River, Maria—Run Around Sue and a Coke, Raol—Hound Dog and a Root Beer. When the bills came, how much did each guy pay for himself and his date?

   Tim and Rhonda ____________

   Raol and Maria ____________

2. All four decided to "do deserts," so they ordered a Hot Fudge Banana Extravaganza and a Strawberry Sundae and said they wanted four spoons "to share." They instructed "Be-Bop Bill" (or so he said) to add half the cost to each of their bills. What did each guy now owe?

   Tim ____________

   Raol ____________

3. The final bill for each couple included a 6.00 % sales tax charge. Add this to each bill and calculate the final bills for both Tim and Raol.

   Tim ____________

   Raol ____________

4. An appropriate tip for places like Little Anthony's Diner is between 10 percent and 15 percent. Tim and Raol decided to be generous and leave 15 percent. How much did each of the boys leave on the table? Round off to the nearest quarter.

   Tim ____________

   Raol ____________

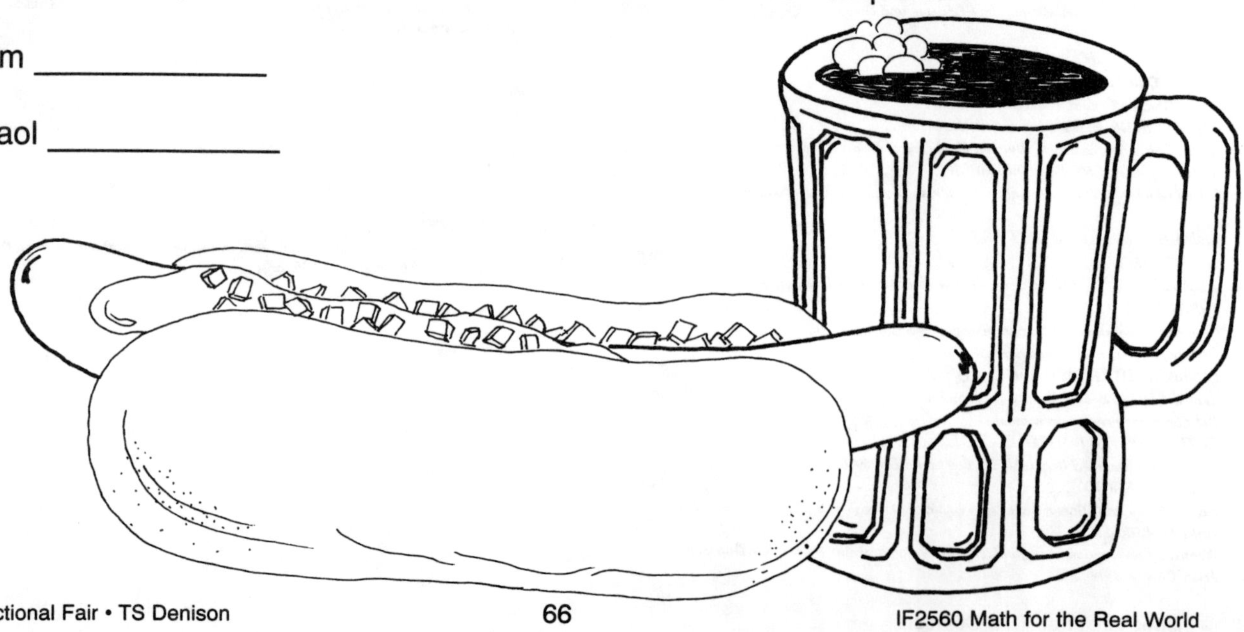

5. Gregg Anthony was a "child of the '60s" who literally grew up with Little Anthony's Diner during the "good old days." In a nostalgic moment he took his entire family there for dinner. Mandy, Gregg's wife, ordered a Garden Burger and a Sprite; Todd, his oldest son, had a Cheezie Burger and a Coke; Laurie ordered a C.C. Ryder and a Coke; and Gregg ordered his favorite—Frenchies' Favorite and a Green River to "wash it down." He ordered a Burger and Chips off the Little Be-Bopper menu plus a Coke for little Matthew (who is only four years old). How much was Gregg's bill?

____

6. Once tax was added, what was the total amount that Gregg owed for his trip down Memory Lane to Little Anthony's? ______________

7. Gregg tipped waiters and waitresses 10 percent during the '60s . . . and he still tips them 10 percent. He says that as prices rise, so does the amount of money that he gives for tips. Based on this philosophy, how much should he leave on the tip tray when the waitress skates up to the table with his change for the bill? Again, round off to the nearest quarter. ______________

8. Now it is your turn to "build your own." Assume that you and a friend, or you and several friends, or you and your family go to Little Anthony's Diner for a night of the '60s. Look at the menu and decide what sounds good. Then order for you and your friend(s) or family. List what you think each person would order. Record and identify each order in the space below and calculate the total bill. Then add the appropriate 6.00% sales tax. Leave a tip of 15% and show calculations for the total bill. Then decide if it will be one bill, or if you want it to be divided (and if so, how). Be prepared to share your menu choices and calculations with other members of your class.

____

____

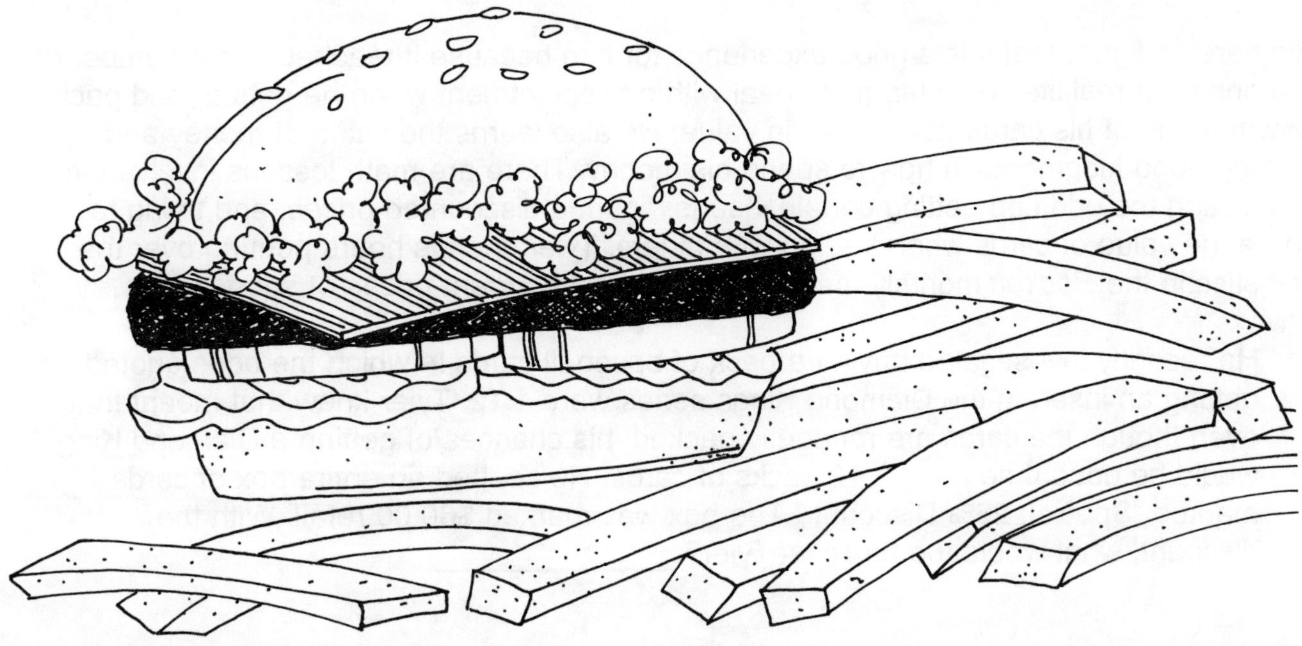

Name ____________________

Tyler's Sports Cards

Tyler collects baseball cards. Does he ever collect baseball cards! And he also collects basketball cards and football cards, too. Since he started collecting three years ago, he has amassed a remarkable collection of cards, some of which are worth a considerable sum of money. In fact, that is why Tyler says he is so "into it." What started out as just an inexpensive hobby to collect the cards of some of his favorite St. Louis Cardinals has grown into a real-life experience worth hundreds, perhaps even thousands of dollars.

Tyler spends most of his allowance on cards, and he also spends much of the other money he earns doing odd jobs around the house plus the money he gets for birthdays, Christmas, and making the honor roll at Coronado Middle School.

Tyler likes to trade cards, too. He trades with his friends and even sometimes with the owner of Mountain View Sports Cards where he buys most of his cards. One of his favorite things to do on weekends is to go to a sports card show where many dealers gather, usually in a large shopping mall.

His parents figure that it is a good experience for him because it teaches him a number of lessons from real life. Tyler learns to deal with disappointment when he gets a "bad pack" or when one of his cards goes down in value. He also learns the value of money and making good judgments in how to spend his money. There are math lessons involved in calculating the odds on getting certain inserts, figuring discounted prices, and trying to figure the value of cards when trading with others. Tyler spends hours pouring over the statistics in the *Beckett* monthly magazines that relate current prices of sports cards.

1. He recently investigated buying a pack of baseball cards in which the odds quoted on getting an insert in the Diamond Kings series were 1:12. Tyler knew that meant that even though the cards are randomly packed, his chances of getting a Diamond King would be good if he bought 12 packs of cards. He spotted an entire box of cards marked "Special: 25% Discount." The box was marked $60.00 retail. With the discount, what would the box cost Tyler? ______________

2. Because Tyler joined the Mountain View Buyer's Club, he gets an additional 10% discount anytime he buys anything at Mountain View. With this additional discount, what is the total cost of the box of cards to Tyler? ____________

3. If sales tax is 6.25%, what is the final cost of the box of cards to Tyler?

____

4. If there are 24 packs in the box, how many Diamond Kings should Tyler expect in his box of cards? ____________

5. Tyler looked at these odds for getting special inserts on the back of a box of basketball cards: Dish and Swish 1:4, Rim Rockers 1:12, Block Party 1:8, Jam City 1:48.

   If Tyler decides to buy a box of these cards, how many of each of the above inserts can he expect to get if there are 32 packs in the box?

   Dish and Swish ____________ Rim Rockers ____________
   Block Party ____________ Jam City ____________

6. On the back of a box of baseball cards Tyler read the following odds about getting inserts in the series: Diamond Kings 1:6; Spirit of the Game 1:12; Heritage Series 1:24; Press Proof 1:36.

   If Tyler decides to buy a box of these cards, how many of each of the above inserts can he expect to get if there are 36 packs per box?

   Diamond King ____________ Heritage Series ____________
   Spirit of the Game ____________ Press Proof ____________

7. One of Tyler's favorite insert series is a set of eight basketball cards called Pan for Gold. He already has two cards, but still needs the other six cards to complete the set. If the odds on getting a Pan for Gold card are 1:12, how many boxes would Tyler have to buy to expect to get the remaining six cards if there are 24 packs per box?

   ____

8. Tyler began doing some calculating to see if pursuing the Pan for Gold set would be a good investment. If a box of cards costs $42.00 (but remember that Tyler will still get his discount), how much will it cost him to have a pretty good chance at getting the remainder of the cards in the set? (Don't forget the 6.25% sales tax.)

   ____

9. Tyler looked in his *Beckett Basketball Monthly* and found the following current values for the cards in the Pan for Gold set: Michael Jordan—$40, Shawn Kemp—$32, Hakeem Olajuwon—$28, Jason Kidd—$20, Anfernee Hardaway—$20, Grant Hill—$15, Scottie Pippen—$12, Clyde Drexler—$8. What is the total value of the individual Pan for Gold set of cards? ______________

   Would Tyler be wise to pursue buying the packs of cards to try to complete the set, or is it a bad investment? ____

   ____

10. At a recent sports card show in the Foothills Mall, Tyler noticed an ad posted by one of the vendors that said: 1996 Score Dream Team—Individual Cards Offered at 30% Below *Beckett*.

    Tyler thumbed through his *Beckett Baseball Monthly* and found these prices: Cal Ripken—$30, Ken Griffey, Jr.—$30, Frank Thomas—$30, Kenny Lofton—$10, Jeff Bagwell—$10, Chipper Jones—$12, and Barry Bonds—$6. These were the only cards from the Dream Team Set the vendor had in his case. If Tyler decides to buy all the cards except the Cal Ripken (which he already has), how much will he have to pay? ______________

11. Tyler has two Emmitt Smith cards from the Playoff Prime Fantasy Team set. He would like to trade one of them to Jason at Mountain View Sports Cards to get other cards in the series which he does not have. Jason (who runs the store) said that he was authorized to give Tyler only 60% of the true value of the card. The Emmitt Smith card is worth $50. If Tyler wants to trade for a Steve Young worth $18 plus a Jerry Rice worth $16 plus a Dan Marino worth $12, how much cash will Tyler need to pay? (Don't forget his 10% discount, but sales tax will not apply in this trading situation.)

    ____

12. The Ultra Home Run Kings set includes the following cards and values: Albert Belle—$25, Dante Bichette—$6, Barry Bonds—$10, Jose Canseco—$8, Juan Gonzalez—$12, Ken Griffey, Jr.—$50, Manny Ramirez—$20, Matt Williams—$8, Mo Vaughn—$8, Tim Salmon—$5. What would the entire series cost at a card show when purchased from a dealer who advertises that he sells at 50 percent below *Beckett*? There is no sales tax at sports card trading shows. ______________

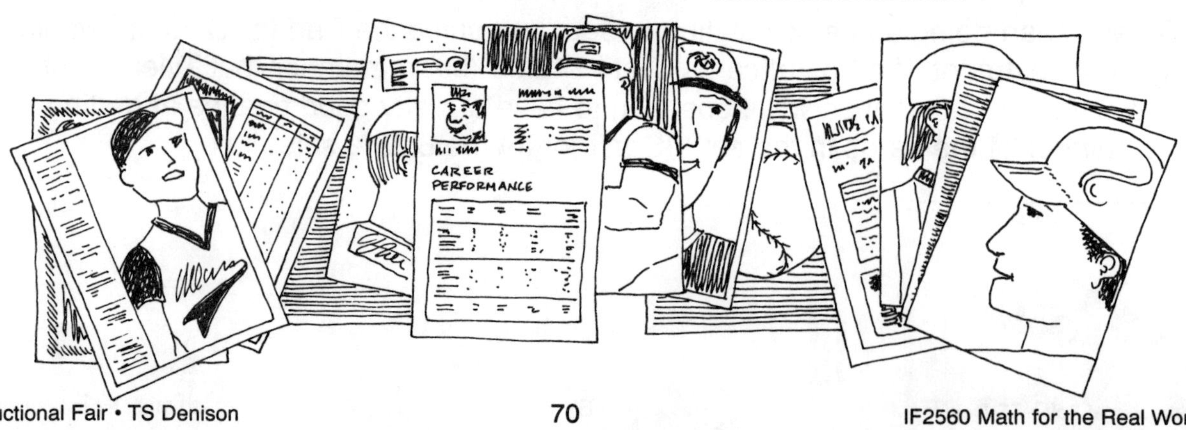

Name ______________________________

Laurie's Laredo

Laurie David has her sights set on a brand new car. She has spent hours scouring several car lots in town, and she has found some pretty nifty automobiles. But none has caught her eye or captured her heart like the brand new flaming red Laredo she spotted the other day! Ever since that day Laurie has been fantasizing about herself behind the wheel of that beautiful red car. There she is ". . . fanning down the freeway with the moon roof pulled back, sunshine pouring in, and the wind blowing through her hair. Add the five-speed stick shift, the four-wheel drive, and Laurie has the freedom to go anywhere anytime! Perhaps she may even buy one of those vanity license plates that might say something like LAURIE DEES. Can it ever get any better than this? WOW!

Laurie has searched high and low, and she has come to the "rational" conclusion that she MUST have that car! No other will do! The only problem she has is money. Laurie really wants a car that costs more money than she can afford. So, Laurie David has decided that she will become really aggressive and get the "best deal" she can get to make the car more affordable to her.

But Laurie quickly finds out that her strategy involves a VERY long road ahead! Car dealers have developed a sometimes deserved reputation for pricing their cars very high in the beginning with the expectation that the potential customer will balk and then feel that he/she is getting a much "better deal" when the price is lowered. Laurie soon realized that she could not afford the car at the price she was given by the salesman. His opening offer was to sell her the car "at sticker price." Laurie looked longingly at that sticker price, but the bottom line that reflected the final cost was torn off. Laurie asked the salesman about this and he told her that he would go to the accounting department to check it out.

While he was gone, Laurie decided to add the figures to verify the true total. But she didn't have a calculator, so she needs your help. Look closely at the "sticker price" and find out the cost of the car Laurie wants to buy.

Manufacturer's Suggested Retail Price			$21,470
Optional Installed Equipment			
Preferred Equipment Package			
Air Conditioning		$950	
Power Windows, Power Locks, and Remote Tailgate Release		550	
Cruise Control		250	
Tilt Steering Column		100	
6-Speaker AM/FM Stereo/Cassette		150	
Dual Visor Vanity Mirrors		10	
Premium Door Trim w/Lights		130	
Cargo Cover and Convenience Net		112	
Aero Roof Rack		190	
Outside Power Mirrors		100	
Front and Rear Intermit Wipers		108	
		$2650	
Discount offered of $500			
Keyless Entry	$210		
Hood Protector	38		
Splash Guards	70		
Destination and Handling		$445	
Dealer-Added Equipment and Services			
Value Protection Package Includes			
Regal Care Paint Sealant			
Regal Care Fabric Sealant			
Regal Care Theft Code		$589	
Total Vehicle Price			

1. The lower right-hand corner of the sticker was torn off. Your task here is to help Laurie by calculating the final sticker price. Record your answer in the space below.

____

2. Laurie had a friend tell her that some people are able to buy cars for at or near the invoice cost to dealers. The invoice cost is the price that the car dealer actually pays for the car. Laurie went back to the dealer and demanded that she be able to see the true invoice cost which the dealership had paid for the car. After some reluctance, the salesman went to his superior, and Laurie was eventually escorted into the Sales Manager's office. There she was shown the original invoice which indicated that the car was purchased for $22,471. What is the difference between what the dealer paid for the car (the invoice price) and the price the dealer is asking for the car?

____

3. One of Laurie's options was to buy her car through a broker. She found one in the Yellow Pages and gave him a call. He claimed he could get her the car she wanted at a cost of "$400 over invoice cost." If Laurie goes this route, how much will the car cost her? ______________________

4. But Laurie still has another problem! She has to find a way to sell the car she is currently driving. Laurie checked at her bank to see how much her car was worth. If she could sell it to the car dealership for the "wholesale price" (it was supposedly worth $15,800), and she buys the new car at "sticker price," how much money would Laurie need to buy the new car? ______________

5. Laurie decided to place an ad in the newspaper. She advertised that she would be willing to sell it for $16,900. One person called, came by to look at the car, and offered her $15,500. If she had taken this offer, then bought her new car through the broker, how much would the car have cost Laurie? ______________

6. Laurie decided to become really aggressive and stormed into the dealership with a "final offer" of $200 over the invoice price for the car. The dealer also gave Laurie $200 more than the wholesale price for her car. How much was the difference?

____

7. If the sales tax is 7.25% of the difference between the agreed-upon price for the new car and the agreed-upon value of Laurie's trade-in, how much tax will Laurie pay?

____

8. Laurie also found out that she would have to pay $530.28 for license plates for her new car. In addition, there was a "processing fee" of $69.00 to cover the cost of the dealership handling all of the paperwork. Adding all of these fees and taxes onto the original agreed-upon price, what was the total amount for which Laurie wrote the check so she could drive away, retract her moon roof, and enjoy her new car?

____

Name ______________________________

House for Sale

ENJOY RELAXING AND QUIET SWEEPING MOUNTAIN VIEWS AND WIDE OPEN SPACES. THIS FORMER MODEL HAS A DRAMATIC TWO-STORY ENTRY. ENJOY PEACEFUL SUNSETS OR EARLY MORNING SUNRISE ON MASTER BEDROOM BALCONY. SCULPTED LANDSCAPE WITH FREE-FORM POOL. ENTERTAINING IS EASY AND STYLISH WITH THIS ENGLISH GARDEN SETTING OFF FRONT ROOM FOR ELEGANT DINING.

Ads like the one above are typical of the push by real estate agencies to sell homes. Oftentimes when people want to sell their home, they hire a real estate firm that is skilled in the art of selling houses. Personnel will offer tips to the seller to make the home look the very best it can look when potential buyers pass through. These agents also know how much money houses of a specific size are selling for in a given neighborhood. They also know all the tricks-of-the-trade that will help the seller to get the most money possible, while at the same time selling the home as quickly as possible.

On the other side of the fence is the home buyer. This individual (or family) usually has a pretty good idea of THE PERFECT HOME. Finding that house is often more difficult than it seems it should be. Once that first hurdle is cleared, the second becomes one of reaching a price that is agreeable to both the buyer and the seller. The real estate agent serves as the liaison between the two. The seller has a price he or she would like to get for the house. The buyer would obviously like to get the home by spending much less money. The offers and counter offers are the very essence of home buying.

There is a lot of pressure involved because there is a lot of money at stake. Once a mutually agreeable price is reached, the next step is borrowing the money. If the buyer has a good credit rating and money in the bank, he or she is in a much better position to bargain with the money lenders on the rate of interest that will be paid for borrowing the money. Rates and conditions vary considerably from one lender to another as they also vary from one period of time to another.

Again, the buyer should "shop around" to find the best deal that will involve the least amount of money. A number as small as one quarter of a percentage point can amount to thousands of dollars over a period of 30 years! The table below shows you how to calculate a monthly house payment that will include both the interest paid for the use of the money (by the buyer) as well as the money paid that helps to reduce the amount of money still owed (principal).

Look carefully at the data presented; then answer the questions that follow.

The Cost of a Mortgage

Monthly principal, interest per $1,000 of mortgage

Mortgage rate	15-year loan	20-year loan	25-year loan	30-year loan
5.000	7.91	6.60	5.85	5.37
5.125	7.97	6.67	5.92	5.44
5.250	8.04	6.74	5.99	5.52
5.375	8.10	6.81	6.07	5.60
5.500	8.17	6.88	6.14	5.68
5.625	8.24	6.95	6.22	5.76
5.750	8.30	7.02	6.29	5.84
5.875	8.37	7.09	6.37	5.92
6.000	8.44	7.16	6.44	6.00
6.125	8.51	6.24	6.52	6.08
6.250	8.57	7.31	6.60	6.16
6.375	8.64	7.38	6.67	6.24
6.500	8.71	7.46	6.75	6.32
6.625	8.78	7.53	6.83	6.40
6.750	8.85	7.60	6.91	6.49
6.875	8.92	7.68	6.99	6.57
7.000	8.99	7.76	7.07	6.66
7.125	9.06	7.83	7.15	6.74
7.250	9.13	7.91	7.23	6.83
7.375	9.20	7.98	7.31	6.91
7.500	9.28	8.06	7.39	7.00
7.625	9.35	8.14	7.48	7.08
7.750	9.42	8.21	7.56	7.17
7.875	9.49	8.29	7.64	7.26
8.000	9.56	8.37	7.72	7.34
8.125	9.63	8.45	7.81	7.43
8.250	9.71	8.53	7.89	7.52
8.375	9.78	8.60	7.97	7.61
8.500	9.85	8.68	8.06	7.69
8.625	9.93	8.76	8.14	7.78
8.750	10.00	8.84	8.23	7.87
8.875	10.07	8.92	8.31	7.96
9.000	10.15	9.00	8.40	8.05
9.125	10.22	9.08	8.48	8.14
9.250	10.30	9.16	8.57	8.23
9.375	10.37	9.24	8.66	8.32
9.500	10.45	9.33	8.74	8.41
9.625	10.52	9.41	8.83	8.50
9.750	10.60	9.49	8.92	8.60
9.875	10.67	9.57	9.00	8.69
10.000	10.75	9.66	9.09	8.78
10.125	10.83	9.74	9.18	8.87
10.250	10.90	9.82	9.27	8.97
10.375	10.98	9.90	9.36	9.08
10.500	11.06	9.99	9.45	9.15

Note: Multiply the cost per $1,000 by the size of the mortgage (in thousands). The result is the monthly payment, including principal and interest. For example, for an $80,000 mortgage for 30 years at 10 percent, multiply 80 x 8.78 = 702.40.

1. Rachel and Aaron Miller have found a split-level they want to buy. The owner wanted $104,995 but, after offer and counter offer, has accepted the Miller's bid to buy for $100,000. If their down payment is 20% of total purchase price (or $20,000), how much is their monthly payment on the $80,000 they need to borrow if they get a loan at 8.0% for 30 years? ______________

2. What will be the total amount of money the Millers will pay during the first year of their loan, including both interest and principal? ______________

3. How much of that amount (question #2) is interest? ____________ How much will the Millers have reduced the principal by the end of the first year?____________ What will be the balance on the principal they still owe as they enter the second year of their loan? _________

4. How much more per month would their payment be if they were to choose a 15-year loan rather than a 30-year loan? ______________

5. If the Millers were able to handle the monthly payments on the 15-year loan, they know they would be saving a lot of money over the entire life of the loan. Aaron and Rachel began punching out numbers on their calculator and were really shocked at the difference! Find out for yourself by taking your answer for #2 and multiplying that number by 30. Then figure out how much Aaron and Rachel would pay for a year if they chose the 15-year loan. Then multiply that number by 15 and subtract the total cost of the 15-year loan from the 30-year loan to find out the difference.

____

6. Rachel and Aaron agreed that the 15-year loan was the much less expensive path to follow. But they also agreed that making the monthly payment would really leave them strapped financially on paying some of their other monthly expenses. So, they have decided on the 30-year loan because it allows them to commit to a lower monthly payment. But before they sign the loan agreement with First Fidelity, Aaron decided to do a little "checking around" to see if he could get a loan elsewhere at a lesser rate.

His diligence and "legwork" paid off as Aaron found that Bank of America would loan him and Rachel the money for 7.5%! Hooray and happy day! A bit of quick figuring led the Millers to realize how much money can be saved by such a seemingly insignificant figure as one-half of a percent! Do your own figuring and see why the Millers were so elated over their good fortune.

How much less per month would their payment be with the 7.5% loan?

____

How much money will they save over the course of an entire year by borrowing the money at 7.5% rather than 8.0%?

____

How much money will the Millers save over the life of the loan (30 years) by borrowing from the Bank of America at 7.5%? ______________

7. How much money will a family pay in interest over the entire life of a 30-year loan if they borrow $100,000 at 7.0%? ______________

8. What is the monthly payment on a $72,000 loan for 25 years at 7.25%?

____

9. How much less money will be spent per month on an 8% loan for 30 years than on a 9% loan of $110,000 for the same amount of time? ______________

10. Choose any loan amount you desire and find out the current interest rate at which banks and lending institutions in your town are loaning money for homes. Using those figures, calculate the monthly payment on a 15-year loan, on a 20-year loan, on a 25-year loan, and on a 30-year loan. Compare your figures with those of other members of your class.____

____

____

Name ______________________

A Fact of Life

Modern medicine combined with breakthroughs in technology have in recent years dramatically extended the life expectancy of both women and men here in the United States. In fact, current data shows a life expectancy of nearly 80 years for a female born today and nearly 75 years for a male. While such statistics are the kind of numbers we like to hear, the fact remains that all human beings die sometime.

Family "breadwinners" often make financial plans to provide for their survivors by buying what is called *life insurance.* It is sort of a no-win gamble. The *insured* buys a set amount of insurance and is charged according to his or her age and the amount of insurance desired. If the insured dies during the period he/she is covered by the policy, the insurance company pays to the insured's survivors the full amount of the policy. If the insured happily lives on, the premium paid is just good news for the insurance company. There *are* insurance policies that provide a sort of investment benefit, but the least expensive life insurance and the type that is the focus of this activity is called *term insurance.*

Term insurance is just like the insurance described above. If the insured lives on past the time period for which the premium is paid, the money spent on the premium is the gain of the insurance company and the loss of the insured. But if the insured dies during the period for which the premium is paid, the insurance company pays the full value of the policy to the insured's benefactors. It is sort of like—you win and the insurance company loses only if you die. If you live, the insurance company is the winner. Ah . . . what peace of mind!

Life insurance companies base their rates on several criteria, but the main factor is age. It would seem reasonable that a policy for a 35-year-old person would be considerably less expensive than the same policy coverage for a 65-year-old person. In recent years it has become common practice among insurance companies to offer separate rate schedules for smokers and nonsmokers. As you might guess, the rates for nonsmokers are considerably cheaper.

Look closely at the premium rates chart on the attached page. Notice that the cost quoted is the cost to the insured *per $1,000 of insurance.*

For example, look at the cost per $1,000 of insurance for a 35-year-old person who does not smoke. For a nonsmoker wanting $100,000 of insurance, the first-year coverage is figured in the following manner:

$1.18 x 100 = $118.00 for the first year coverage.

For a policy coverage between $250,000 and $499,000, this company offers a 10% discount. Thus, our 35-year-old person who now wants $250,000, would pay a premium according to this calculation.

$1.18 x 90% = $1.06

$1.06 x 250 + $265.00 for the first-year coverage

For a policy coverage of $500,000 or more, this company offers a 25% discount. Now the problem for calculating the 35-year-old's first-year premium is found in this manner:

$1.18 x 75% = $0.89

$0.89 x 500 = $445.00 for the first-year premium.

This company offers a reduction in cost coverage based on the following formula. Payments for the second and all subsequent years are the difference between the "Guaranteed Premium" (for your age at the time you renew) and the "Projected Dividend." To illustrate this, refer back to the 35-year-old person with the $100,000 coverage who would pay just $118.00 for the first-year premium. Beyond that, look at the calculations below.

  $264.00 (age 36 premium of $2.64 x 100)
- $140.00 (projected dividend for age 36)

= $124.00 (second-year projected payment)

Now, look carefully at the chart on the next page and answer the questions that follow.

SELECT* (rates for nonsmokers)

AGE	GUARANTEED FIRST-YEAR PREMIUM	RENEWALS: PROJECTED PAYMENT†	RENEWALS: GUARANTEED PREMIUM	RENEWALS: PROJECTED DIVIDEND†
18-30	$1.17	$1.17	$2.50	$1.33
31	1.17	1.17	2.51	1.34
32	1.17	1.17	2.51	1.34
33	1.18	1.18	2.52	1.34
34	1.18	1.18	2.52	1.34
35	1.18	1.18	2.52	1.34
36	1.24	1.24	2.64	1.40
37	1.27	1.27	2.83	1.56
38	1.34	1.34	3.01	1.67
39	1.43	1.43	3.24	1.81
40	1.49	1.49	3.50	2.01
41	1.59	1.59	3.77	2.18
42	1.73	1.73	4.06	2.33
43	1.87	1.87	4.36	2.49
44	2.02	2.02	4.70	2.68
45	2.18	2.18	5.09	2.91
46‡	2.36	2.32	5.51	3.19
47	2.57	2.53	5.95	3.42
48	2.79	2.69	6.45	3.76
49	3.06	2.89	7.00	4.11
50	3.35	3.12	7.63	4.51
51	3.66	3.38	8.38	5.00
52	4.05	3.68	9.23	5.55
53	4.50	4.06	10.23	6.17
54	5.02	4.45	11.37	6.92
55	5.63	4.96	12.67	7.71
56	5.22	5.60	14.14	8.54
57	5.73	6.29	15.63	9.34
58	6.31	7.11	17.32	10.21
59	7.06	8.13	19.23	11.10
60	6.90	6.83	18.14	11.31

STANDARD* (rates for smokers)

AGE	GUARANTEED FIRST-YEAR PREMIUM	RENEWALS: PROJECTED PAYMENT†	RENEWALS: GUARANTEED PREMIUM	RENEWALS: PROJECTED DIVIDEND†
18	$1.88	$1.88	$2.94	$1.06
19	1.88	1.88	3.04	1.16
20	1.88	1.88	3.08	1.20
21	1.88	1.88	3.10	1.22
22-30	1.88	1.88	3.11	1.23
31	1.88	1.88	3.12	1.24
32	1.88	1.88	3.22	1.34
33	1.89	1.89	3.40	1.51
34	1.89	1.89	3.60	1.71
35	1.89	1.89	3.83	1.94
36	1.95	1.95	4.12	2.17
37	2.03	2.03	4.49	2.46
38	2.12	2.12	4.89	2.77
39	2.25	2.25	5.36	3.11
40	2.39	2.39	5.89	3.50
41	2.58	2.58	6.46	3.88
42	2.79	2.79	7.09	4.30
43	3.03	3.03	7.78	4.75
44	3.31	3.31	8.51	5.20
45	3.59	3.59	9.30	5.71
46‡	3.98	3.93	10.12	6.19
47	4.40	4.26	11.01	6.75
48	4.85	4.64	11.98	7.34
49	5.43	5.12	13.03	7.91
50	6.06	5.61	14.23	8.62
51	6.75	6.17	15.60	9.43
52	7.56	6.82	17.17	10.35
53	8.47	7.54	18.98	11.44
54	9.50	8.35	20.96	12.61
55	10.66	9.27	23.14	13.87
56	10.32	10.50	25.54	15.04
57	11.40	11.81	27.97	16.16
58	12.60	13.32	30.57	17.25
59	13.94	14.99	33.41	18.42
60	15.38	13.23	31.97	18.74

* SELECT rates are available to people who have not smoked during the past 12 months. STANDARD rates are available to smokers.

† PROJECTED PAYMENTS for RENEWALS are equal to the GUARANTEED PREMIUM less the PROJECTED DIVIDENDS. PROJECTED DIVIDENDS are based on current dividend scale and cannot be guaranteed for the future.

‡ Increased dividends, due to projected improvements in mortality experience at older ages, cause the PROJECTED RENEWAL PAYMENTS at age 46 and older to be lower than the GUARANTEED FIRST-YEAR PREMIUMS at these same ages.

1. How much would it cost a 41-year-old nonsmoker to buy $100,000 during the first year of the policy? ______________

2. What would the same policy cost if the insured smoked? ______________

3. Assume a 50-year-old nonsmoker wants to buy $200,000 worth of life insurance. What will the annual rate for this policy be after the first year? ______________

4. How much more is the annual premium for a 40-year-old smoker who buys a $250,000 life insurance policy than the same policy purchased by a 40-year-old nonsmoker?

____

5. Comparing the same annual rates, how much older can a nonsmoker be and pay almost the same first-year premium as a 21-year-old pays who *does* smoke?
   ____

6. How much does a policy for $500,000 cost a nonsmoking 46-year-old person during the first year of the policy coverage *per quarter*? ______________

7. The insurance company adds $2 per payment for anyone desiring to pay either semiannually or quarterly. This being the case, how much do quarterly payments cost someone who does not smoke and just bought a policy for $400,000 in celebration of her thirty-seventh birthday? ______________

8. Someone who buys a $1,000,000 policy gets a 25% discount off the guaranteed first-year premium rates. If that person is 56 years of age and does not smoke, how much is the annual premium rate?
   ____

9. What is the second-year annual premium rate for a 60-year-old nonsmoker who buys a policy for $750,000? ______________

10. Create a formula that can be used to express how the second-year projected payment is determined. ______________________________
    ____

11. A 53-year-old man buys $1,000,000 in term life insurance. What will he have to pay next year? The man has never smoked in his life. ______________

Teacher Notes

THE SPORTS PAGE

It is nearly impossible to pick up a newspaper and turn to the sports page and NOT be inundated with math! Statistics literally fill the sports page. It is really what sports are all about! The measure of a team's success against other teams, the performance of an individual player compared to the performances of other individuals, and the statistics that filter down through the ranks are what make the sports page among the most popular sections in the daily newspaper to many of us.

Middle school students are sports fans, too! Those who play the sports are obviously interested in the performances of the college and professional teams that play "their sports," but there are a lot of students (and adults, too) who do not actually participate, but enjoy following the action down on the field. The statistics that fill any daily newspaper require a certain amount of knowledge and understanding to make them meaningful. That's the whole idea behind this chapter.

Following Baseball's Best Hitters

Problems in this activity focus on students calculating the batting averages of some of the major league's leading hitters. Encourage students to use calculators and go over the example with them to ensure their correct technique. They may well be familiar with figuring percentages, but expressing batting averages to the third place after the decimal may be a new experience for them.

It Pays to Play!

Students get involved with some really big numbers as they analyze the salaries of the highest paid players in professional baseball. Included are several problems that relate to the annual payroll of the New York Yankees, the most expensive franchise in major league baseball. Providing students with calculators for this activity is suggested.

Are They Worth It?

Aside from the math problems in this activity, the theme of the problems questions the value of the absurdity of huge salaries paid to today's superstars. Just how much does Deion Sanders get for each minute of football he plays? How much does Greg Maddux get each time he walks to the mound to start a game for the Atlanta Braves? How do these and other questions rank with the salary of the president of the United States? When discussing this question, it is suggested that the salaries of other important Americans be discussed as well, i.e., the annual salaries of teachers, doctors, lawyers, firefighters, police, etc. To add realism to the issue, assign various students the task of researching different salaries. When students return with their findings, have them compare those yearly and daily earnings to those of some of the superstars.

Those Amazing Bulls!

Suggest that students use calculators for this activity that focuses on calculating winning percentages of the record year of the Chicago Bulls professional basketball team (1995-96). In addition to practice in figuring winning percentages, students learn how teams are seeded in the NBA Playoff Championship Series and the importance of the home court advantage.

Predicting Winners

There are no correct or incorrect answers for this activity as students predict winners during a typical weekend of football in the NFL. Their only source of information is current records plus the knowledge of which teams are playing at home. To add realism, it might be fun to bring in a current schedule of weekend play of whatever seasonal professional sport is in play and have students predict winners based on the same kinds of information. Then on the Monday after the weekend, they can compare their predictions with the actual outcomes of the games.

Batter Up!

The purpose of this activity is to give students an opportunity to solve an equation without using an equation. They combine reasoning with exchanging unknowns or use the trial-and-error method to figure out the cost of the bat and the glove. There can be a variety of lines of reasoning that can lead to the current solution. Be certain to investigate the thinking behind each correct solution.

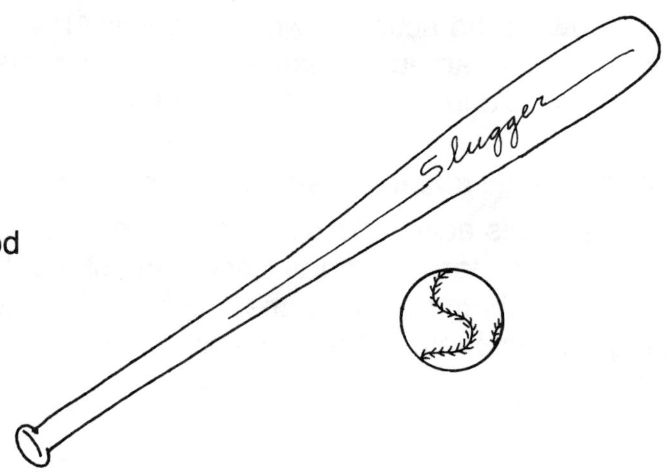

Big Time Franchise—Big Time Buck$

Big numbers are the focus of this activity as students calculate rising and falling percentages of major league sports franchises. While a calculator is not a necessary tool for the completion of this activity, it is suggested as a good activity for calculator practice.

Take Me Out to the Crowd

Students become aware of the reality of some of the big sources of revenue associated with an extravaganza as large as the Super Bowl. Problems investigating the income from the sale of hot dogs, soft drinks, game programs, T-shirts, and parking bring home the realism of how really big the income from such a game can be.

Buy Me Some Peanuts and Crackerjacks!

This activity is another problem in dealing with the exchange of unknowns. Students find the selling price of various items found at a concession stand without knowing the price of any given item. Be certain to have each student who has a different line of reasoning explain his/her thinking in arriving at the correct solution as several are possible.

$uper Commercials

Students are introduced to the high cost of advertising as they get a taste of how expensive it is to advertise during the Super Bowl. It is simple calculator math, but it is math in the real world and will probably be new information to students who have no idea how costly television advertising really is!

Torch of Peace

The activity combines math and geography as students trace the path of moving the Olympic Torch across the United States to launch the 1996 Olympic Games in Atlanta, Georgia. Initiate a discussion with students on reasons for the torch passing through so many states and covering such an extended path across the United States. Discuss also the various methods, other than runners, used to carry the torch.

The Fastest Man on Earth

All problems relate to the 100-meter dash competition in the Olympic Games. Students figure the speed of various athletes, calculating both the number of meters and feet traveled per second.

Marathon Man

All problems focus on the running of the marathon, a distance of 26 miles, 385 yards or 42.19 kilometers. Before students work the problems, discuss with them the length of the marathon. What can they relate to in their own area that would be a distance comparable to the marathon? Perhaps the distance to a nearby town or landmark will help to make the distance more realistic.

Name ____________________

Following Baseball's Best Hitters

Among the more important and most read statistics found on the sports page are the current batting averages of major league baseball players. Such numbers are published daily and help the fans to keep tabs on the success (or failures) of their heroes. Batting averages are calculated simply by dividing the number of times a player has hit the ball and safely reached a base by the number of times he has batted. The division is carried out to the fourth decimal place and then rounded back to the nearest thousandth.

Walks are not counted as official at-bats because the batter has not really done anything to get himself onto first base other than exercising good judgment in not swinging at the four pitches that were thrown to him outside the strike zone. If the batter reaches first base because one of the defenders commits an error, the scorekeeper assesses the batter an at-bat but does not credit the batter with a hit.

While there are a number of statistics that help to identify the playing proficiency of major league baseball players, batting average is among the most important. A good day at the plate can raise a player's batting average, while a bad day can lower the average. Since a player's batting average follows him throughout the season, the later into the season and the more at-bats he has, the less a good day or a bad day affects his batting average.

For example, if a batter gets one hit out of three trips to the plate, it becomes a case of dividing the one hit by the three at-bats to yield a batting average of .333. If another batter gets one hit and two walks out of four trips to the plate, the scorekeeper records an effort of one hit in two at-bats (because walks do not count as at-bats) or an average for the day of .500.

Use your calculator to figure the batting averages in the problems below.

1. During a recent game with the Detroit Tigers, Albert Belle of the Cleveland Indians had two hits in five trips to the plate. What was his batting average for that particular day? ____________

2. Ryne Sandberg of the Chicago Cubs recently was said to have batted .385 during a series with the Houston Astros. If Sandberg recorded 13 at-bats, how many hits did he get? _______________

3. Early in a recent baseball season, the leading batters in the American League were published in the morning newspaper. Calculate the batting average of each; then rank order the top 10 hitters according to their batting averages.

Player, Team	# of Games	AB	Hits	Avg.	Rank
Cordova, Minn	27	106	36	____	____
Alomar, Balt	37	149	56	____	____
Belle, Clev	35	134	46	____	____
Boggs, NY	32	127	43	____	____
Giambi, Oak	28	141	51	____	____
Thomas, Chi	37	138	47	____	____
Knoblauch, Minn	29	112	38	____	____
O'Neill, NY	34	130	49	____	____
Brosius, Oak	28	94	33	____	____
Molitor, Minn	35	143	49	____	____

4. The National League leaders were also listed. Calculate their batting averages and rank order them too!

Player, Team	# of Games	AB	Hits	Avg.	Rank
Piazza, LA	37	146	51	____	____
McGriff, Atl	38	147	52	____	____
Gwynn, SD	29	111	42	____	____
Bagwell, Hou	39	141	50	____	____
Williams, SF	37	141	48	____	____
Grace, Chi	36	146	52	____	____
Grudzielanek, Mtl	39	176	64	____	____
Segui, Mtl	35	132	45	____	____
Joyner, SD	35	127	46	____	____
Rodriguez, Mtl	34	133	47	____	____

5. During a recent year, Tony Gwynn of the San Diego Padres batted .358 and had 490 trips to the plate. How many hits did he have? _______________

6. Frank Thomas of the Chicago White Sox had 174 hits and batted .317. How many at-bats were credited to Thomas for the year? _______________

Name ____________________

It Pays to Play!

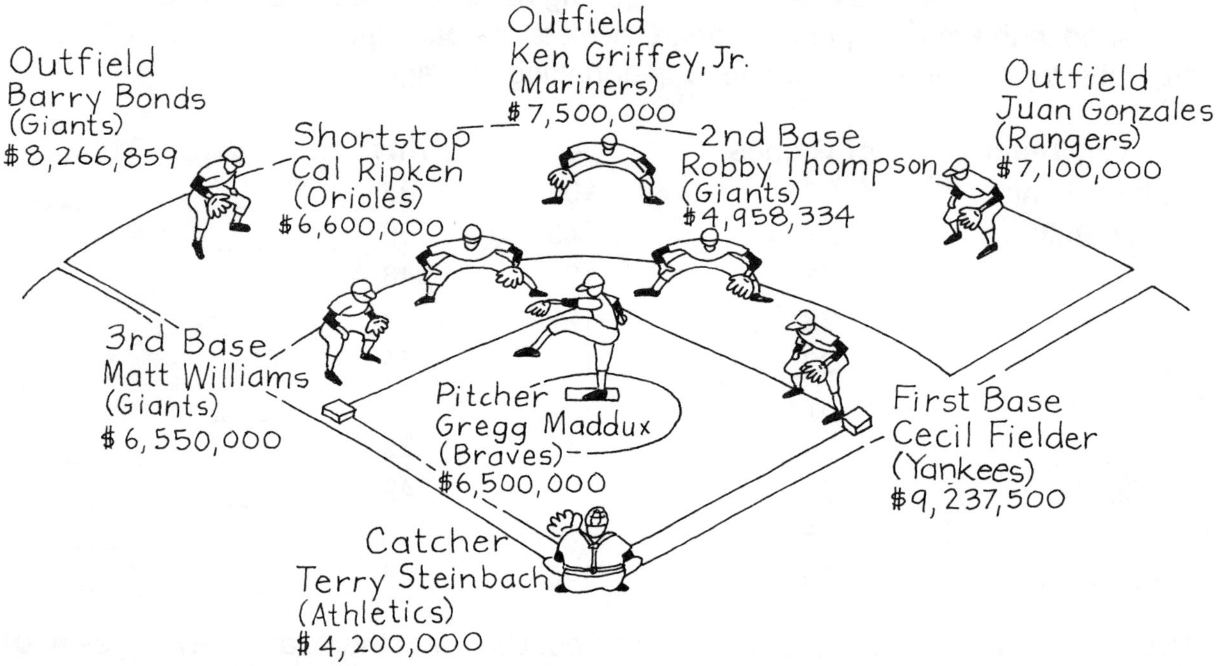

The lineup you see above is the most expensive in baseball! Of course, all the players do not play for the same team. In reality, each man represents the highest paid major league player at each respective position during a recent year. These men have parlayed their talents (with the help of their agents) into earnings that bring them millions of dollars per year. Based on this information, answer the questions that follow.

1. What would the annual payroll be if all these players played for the same team?
   ____

2. Cal Ripken, who broke Lou Gehrig's record for playing in consecutive games in 1995, plays every day. Based on a 162-game season, how much does Cal Ripken get on a per-game basis? ____________

3. How much more per game does the most expensive player get than the "least expensive" of these highly paid professionals? ____________

4. Ruben Sierra of the New York Yankees is the highest paid designated hitter (DH) in the American League. The National League does not use designated hitters. The purpose of the designated hitter is to provide a proficient hitter in place of the pitcher. Pitchers are notorious for being poor hitters as they spend their time in practice working on their pitching. Rubin Sierra earns $6,200,000 per year and his main assignment is just to bat as the designated hitter. If Sierra bats an average of 4.2 times per game for the entire season, how much money did he make per at-bat? Round your answer to the nearest hundred dollars. ____________

5. The New York Yankees had the highest payroll in baseball in a recent year. Yankee owner George Steinbrenner's philosophy is that the quickest way to become a winner is simply to go out and buy the best players available. As a result, the Yankee payroll is larger than the payroll of any other major league team. Just how large that payroll really is becomes the focus of this part of the activity. Below is the Yankee roster from a recent year and the salaries made by each player on the team. Your task is simply to add them up. It would be wise to have a calculator with a printer so you can check your addition.

Ruben Sierra, of/dh	$6,200,000
Paul O'Neill, of	5,300,000
Kenny Rogers, p	5,000,000
David Cone, p	4,666,666
Melido Perez, p (DL*)	4,650,000
John Wetteland, p	4,000,000
Bernie Williams, of	3,000,000
Tino Martinez, 1b	2,300,000
Joe Giradi, c	2,250,000
Wade Boggs, 3b	2,000,000
Jimmy Key, p	1,750,000
Tony Fernandez, ss (DL)	1,500,000
Tim Raines, of (DL)	1,450,000
Jim Leyritz, c	1,330,579
Scott Kamieniecki p (DL)	1,100,000
Pat Kelly, 2b (DL)	900,000
Jeff Nelson, p	860,000
Dwight Gooden, p	850,000
Bob Wickman, p	800,000
Mariano Duncan, 2b	725,000
Steve Howe, p	500,000
Gerald Williams, of	200,000
Andy Pettitte, p	150,000
Mariano Rivera, p	131,125
Derek Jeter, ss	120,000
Mark Hutton, p	119,000
Robert Esnhoorn, ss	119,000
Andy Fox, 2b	109,000
Matt Luke, of	109,000
TOTAL PAYROLL	

*DL = Disabled List

6. What is the average salary of a New York Yankee for the year based on the payroll from the previous page? (Include both active players as well as the five players on the Disabled List.) Round your answer to the nearest dollar. ______________

7. If the New York Yankees play 162 games during the entire season, how much does it cost Yankee owner George Steinbrenner to "put the Yankees on the field" each time they play? ______________

8. While the New York Yankee payroll was the largest for the year, the World Champion Atlanta Braves were not far behind. The Braves are owned by Ted Turner, who also believes that the way to success in major league baseball comes with paying large salaries to the best players money can buy. The Braves' team payroll for the year was $47,930,000. How much did it cost Ted Turner to field the Atlanta Braves each game?

____

9. The Montreal Expos have the least expensive payroll in major league baseball with a total team payroll of $15,410,500. If there are 26 players included on the payroll, how much more is the average salary of the most expensive team in baseball (the Yankees) than that of the least expensive team in baseball (the Expos)?

____

10. The average salary of a major league baseball player in 1976 was $52,300. In 1996 (20 years later), the average salary was $1,172,734! First, find the difference between the two; then calculate the total percentage increase between the 1976 salary and the 1996 salary.

    Difference between average salary in 1976 and 1996: ________________

    Total percentage increase between 1976 and 1996: ________________

    Do you have any closing comments to make about the high cost of major league baseball? If so, jot them down on the back of this page and share them with other members of your class.

Are They Worth It?

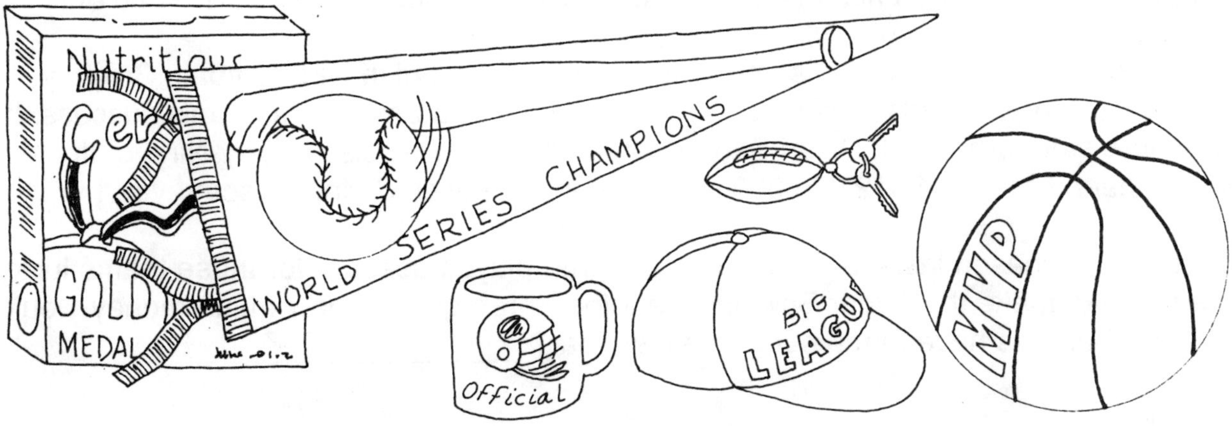

We hear a lot of talk these days about the highly overpaid athletes who take us far from the "good old days" when professional athletes played the game for their love of the game and the desire to win. The salaries they were paid were nice to earn, but they were not really the reason for playing. These days the call of "take me out to the ballgame" is probably not just an appeal of children to Dad, but also the cry of Dad's wallet! It might be a good idea for Dad to stop by the bank on his way to the game. The price of a trip to the ballpark these days, according to a recent survey, is just a shade over $104 if you are going to a major league baseball game.

The **average** price of a ticket to watch the less-than-successful Arizona Cardinals play at home in Sun Devil Stadium is $67.50! Ticket prices range from $20 to $200, and with the stadium seating 73,521, a sell-out crowd becomes a real bonanza for the team owner, even though the Cardinals seldom win. Those who can afford the cost of admission should also add on the cost of another $6.00 for parking. Throw in a few $4.00 hot dogs and $3.00 soft drinks, and you can easily see what an expensive day it becomes for poor old Dad!

What causes it all? Some people feel it is the outright greed of the team owners in their desire to "sting the fan" in the stands for just as much as possible. Others have a different explanation. Some fans feel the real culprit is found in the salaries paid to the players (and the coaches) down on the field. In their desire to win games and divisional championships, league championships and world championships, many of the owners have become far too extravagant with the amount of money they are paying their players. The logic follows that they attempt to attract the better players available from other teams (free agents) and then try to keep the best ones they already have by catering to their high salary demands.

This creates a domino effect among the other teams with the big checkbooks and a desire to do the same thing. The end result becomes a case of passing the bill along to the fans themselves in the form of higher prices for tickets. Championships and television coverage and other increments amount to millions of dollars annually. But it is never enough! To meet the annual payrolls required these days, owners go after the loyal fans themselves by raising the cost of the price of admission. How much does all this cost? Read on!

The Dallas Cowboys' franchise is one of the most successful in NFL football. Cowboy owner Jerry Jones spends millions of dollars every year to ensure his team's success. In fact, to lure star cornerback Deion Sanders from the San Francisco Forty-Niners, Jones paid Sanders a reported $12,078,000 in salary and bonuses during a recent year!

1. If the NFL season lasts for 16 games, and the "playoff season" for those teams lucky enough to make the Super Bowl lasts another three games, how much money would Deion Sanders have made per game last year? ______________

2. All NFL games involve four quarters of football, each lasting 15 minutes. Since Deion Sanders is a cornerback, he is only on the field when the opposing team has the football. If we were to assume that he was on the field for half the game (30 minutes), how much would he be making for every minute during the time he is on the field? ______________

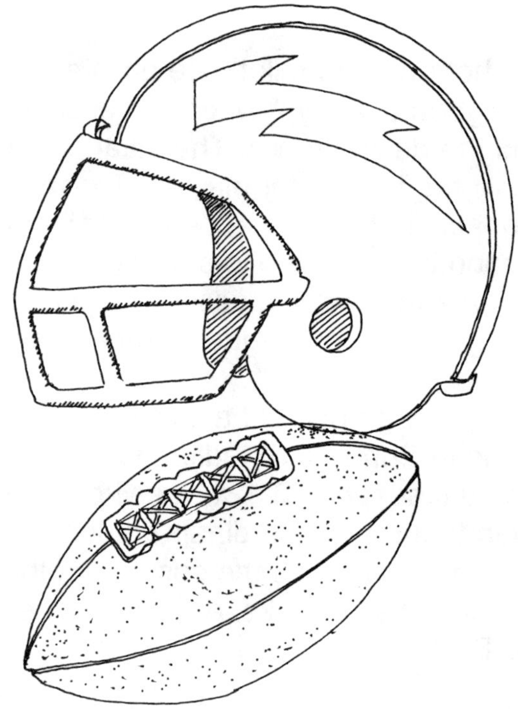

3. During this same year it is said that Jerry Jones spent $61.1 million on salaries and bonuses paid to his players. This being the case, how much did each minute of football cost the Dallas Cowboys for the entire year? (For this problem, assume the Cowboys advanced all the way to the Super Bowl.) ______________

4. If Jones paid $39.3 million in signing bonuses and another $20.6 million in base salaries, how many players are on the roster if the average cost per player was $767,948? ______________

5. The annual salary of the president of the United States is $200,000. Deion Sanders gets over $12 million dollars for playing football! Develop some sort of comparison on your own that shows the relationship of these two salaries. You might use a bar graph or any other kind of mathematical presentation of your desire.

What does this say about our values system here in America? Add your own personal comments in the space below, and share your thoughts with other members of your class.

____

____

____

6. Greg Maddux of the World Champion Atlanta Braves is one of the best pitchers in baseball. He recently signed a contract for an annual salary of $6,500,000. They call him the Six-Million-Dollar Man! The major league baseball season lasts for 162 games. But a pitcher only pitches every fourth or fifth day, depending on the rotation of his team. If Greg Maddux starts 33 games during the season, how much money is he making for each start? ______________

7. A starter in today's major league baseball game averages only about 100 pitches. Beyond that the arm begins to show signs of fatigue, and oftentimes a relief pitcher is brought in to finish the game. Based on the salary information on Greg Maddux you were given above, how much money is Maddux making per pitch? ______________

Name ____________________

Those Amazing Bulls!

During the 1995-1996 NBA season, the Chicago Bulls became the winningest team in NBA history by winning 72 of their 82 official games. This amazing accomplishment placed the Bulls among the greatest teams ever! The Bulls' remarkable season was propelled by the return of superstar Michael Jordan, who had retired two years earlier in an attempt to launch a career in professional baseball. Jordan returned to the Bulls late in the 1994-1995 season and was going full-throttle again by the beginning of the 1995-1996 season.

1. In professional basketball the team standings are recorded and monitored by fans all over the world by the team's winning percentage of all games played. The percentage is calculated by dividing the number of wins by the number of games played. The result is rounded off to three decimal places. Based on that information, what was the Bulls' winning percentage? ______________

2. The Bulls won 39 games on their home court in Chicago and lost only two! What was the Bulls' winning percentage at home? ______________

3. The Bulls' on-the-road record was 33 wins and eight losses. What was their winning percentage while on the road? ______________

4. The next best team in the Eastern Conference was the Orlando Magic, which won 60 of its games. What was the winning percentage? ______________ How many percentage points did the team finish behind the Bulls? ______________

5. The worst team in the Eastern Conference was the Philadelphia 76ers, which won only 18 of its games. What was the winning percentage? _______________ How many percentage points did the 76ers finish behind the Bulls? _______________

6. Another way fans have of identifying a team's success is to calculate the number of games a team is behind the league's leading team. As established in problem #4, the Orlando Magic finished with the second best record in the Eastern Conference. How many games behind the Bulls did the Orlando Magic finish? _______________

7. The Philadelphia 76ers had the worst record in the Eastern Conference. How many games were they behind the Bulls when the season ended? _______________

8. The Seattle Supersonics won the Western Conference Championship in the same year with a record of 64 wins and 18 losses. How many percentage points were they from equaling the pace set by the Bulls?

____

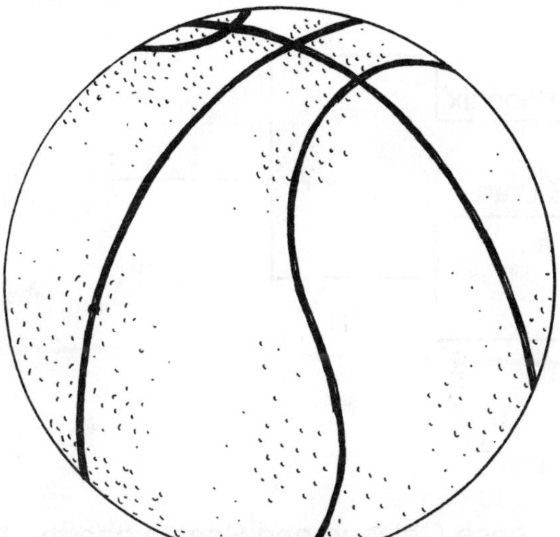

9. The team with the worst record in the NBA was the expansion team from Vancouver, which won only 15 games. Out of each ten games, how many did they win? (Round off to the nearest whole game.)

____

10. The Miami Heat was the last team to make the playoffs in the Eastern Conference. The Heat won only 42 games, but yet it was good enough to make the playoffs. What was the Heat's winning percentage? _______________

On the next page is the format for the 1996 NBA Championship Series Playoffs. Notice that eight teams from each conference make the playoff series. Look carefully at the format and answer the questions that follow.

Western Conference

Eastern Conference

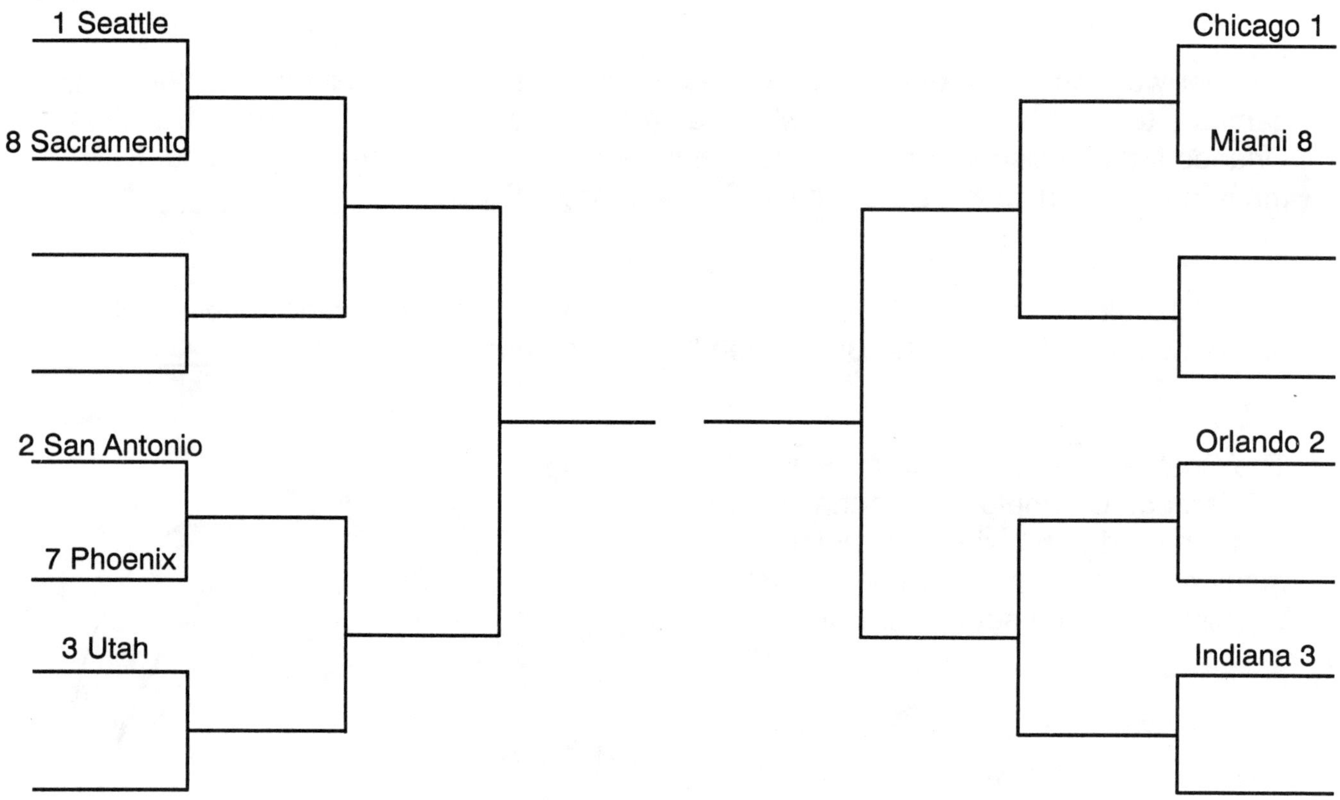

11. Since Chicago and Seattle had the best records in their respective conferences, both received the #1 ranking going into the playoffs. Each team then played the team with the worst record in the conference that still made the playoff series. The teams with the next best records in the two conferences were the San Antonio Spurs in the Western Conference and the Orlando Magic in the Eastern Conference. They both received the #2 rankings and were placed in the bottom brackets. The teams they played were the teams from each conference with the next-to-worst records (#7) to still make the playoffs. This is the plan the NBA uses to pit the better teams against the lesser teams during the earlier rounds of the playoffs, the hope being to build toward a dramatic finish between the very best teams.

    The Utah Jazz (in the Western Conference) and the Indiana Pacers (in the Eastern Conference) received the third seeds. You should begin to see the NBA's logic in the way it creates the brackets. Each first-round game features two teams with a combined seeding totaling 9 points (i.e., 1 + 8, etc.) Label the numerical seedings for the rest of the chart above showing the manner in which the teams were seeded according to their final win/loss record.

Name ______________________

Predicting Winners

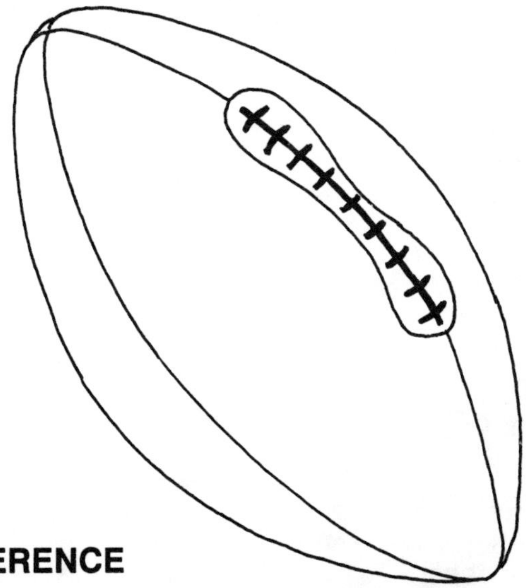

Below are the standings after Week #9 of a recent season in the NFL. Look at the records to see "Who's been winning?" and "Who's been losing?" Beneath the standings is the schedule for the games to be played during Week #10.

Based on the current team records and your superior knowledge of NFL football, predict the team you think will win each game. One additional factor to bear in mind is that the home field advantage is usually valued at three points.

AMERICAN CONFERENCE

EAST	**Wins**	**Losses**	**Ties**	**Percentage**
Buffalo	6	3	0	.667
Miami	6	3	0	.667
Indianapolis	5	4	0	.555
New England	3	6	0	.333
New York Jets	2	8	0	.200
CENTRAL				
Pittsburgh	5	4	0	.555
Cleveland	4	5	0	.444
Houston	4	5	0	.444
Cincinnati	3	6	0	.333
Jacksonville	3	6	0	.333
WEST				
Kansas City	8	1	0	.888
Oakland	7	2	0	.777
Denver	5	4	0	.555
San Diego	4	5	0	.444
Seattle	3	6	0	.333

NATIONAL CONFERENCE

EAST	Wins	Losses	Ties	Percentage
Dallas	8	1	0	.888
Philadelphia	5	4	0	.555
New York Giants	3	6	0	.333
Arizona	3	6	0	.333
Washington	3	7	0	.300
CENTRAL				
Chicago	6	3	0	.666
Green Bay	5	4	0	.555
Tampa Bay	5	4	0	.555
Minnesota	4	5	0	.444
Detroit	3	6	0	.333
WEST				
Atlanta	6	3	0	.666
St. Louis	5	4	0	.555
San Francisco	5	4	0	.556
Carolina	4	5	0	.444
New Orleans	3	6	0	.333

WEEK #10 SCHEDULE

Place the name of the team you predict will win each of the games in the blank space beside each game. Be prepared to defend your choice.

____________________ Indianapolis at New Orleans

____________________ Oakland at New York Giants

____________________ Atlanta at Buffalo

____________________ New England at Miami

____________________ Cincinnati at Houston

____________________ Seattle at Jacksonville

____________________ Carolina at St. Louis

____________________ Tampa Bay at Detroit

____________________ Chicago at Green Bay

____________________ Kansas City at San Diego

____________________ Minnesota at Arizona

____________________ Denver at Philadelphia

____________________ Cleveland at Pittsburgh (Monday Night Football)

Open Dates for New York Jets and Washington Redskins

Name ______________________

Batter Up!

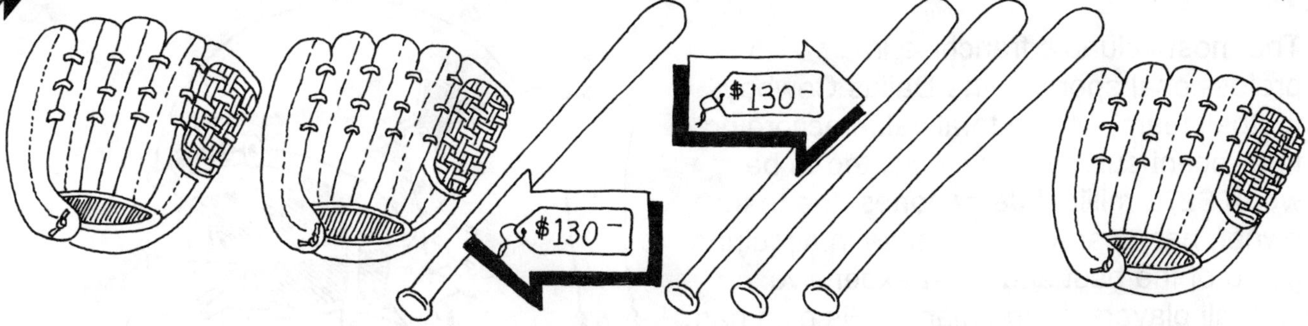

1. Just by looking at the drawings above, which costs more, the bat or the glove? ____________

   Describe the logical reasoning you used to arrive at your conclusion.

   ____

   ____

2. How much would four gloves and two bats cost? ____________

3. How much would six bats and two gloves cost? ____________

4. How much would three gloves and four bats cost? ____________

5. What is the relationship between the price of the bat and the price of the glove?

   ____

   ____

6. How many bats could be purchased for $130? ____________

7. How much does one bat cost? ____________

8. What is the price of a glove? ____________

Name ______________________

Big Time Franchise—Big Time Buck$

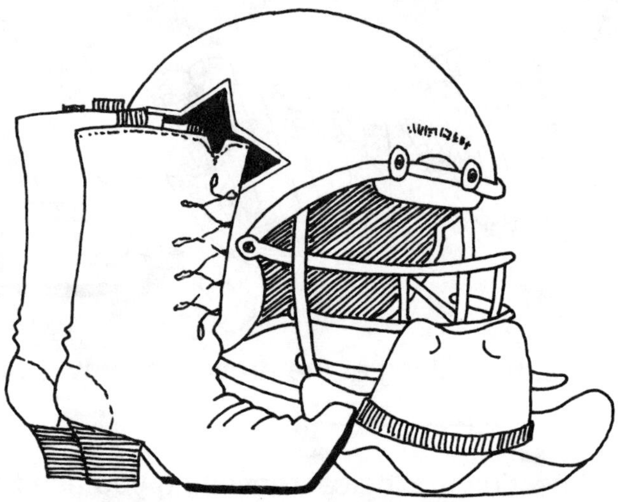

The most valuable franchise in professional sports is the Dallas Cowboys football team, whose total value according to a recent survey was calculated to be worth $272 million! Jerry Jones, the team owner, has spared no expense in acquiring some of the best and most expensive football players on the planet. His plan has paid off because the Cowboys, known affectionately as "America's Team," win football games and football championships.

1. According to the survey, the Dallas Cowboy franchise grew in one year from a projected value of $238 million to $272 million. What percent did Jerry Jones make on his money for the year? (Round to the nearest whole percent.) ______________

2. The next most valuable franchise in the NFL was the Miami Dolphins, whose franchise was estimated to be valued at $214 million. If the franchise was said to have grown 15.1% over the past year, what was the value of the Dolphins franchise last year? ______________

3. During the year of the survey, the team franchise which showed the greatest growth was that of the St. Louis Rams. The increased value was no doubt the result of the team's move from Los Angeles, a city where the Rams had trouble getting fans to come to games, to St. Louis, a city that was starving for a football team and welcomed the Rams with open arms. If the Rams franchise is currently worth $193 million after a growth of 26% last year, what was the value of the franchise during its last year in Los Angeles? ______________

4. In major league baseball, the New York Yankees franchise is considered the most valuable (according to the survey) with a current value of $209 million. Last year the franchise was valued at $185 million. What was the growth rate of the Yankees franchise over the past year? ______________

5. The Montreal Expos franchise is currently valued at $68 million. Last year the franchise was valued at $76 million. What percent did the franchise decline in value over the past year? _______________

6. Last year the Kansas City Royals franchise was valued at $96 million. This year its value declined 16.7%. What is the current value of the Royals franchise?

____

7. The least valuable franchise in professional baseball is that of the Pittsburgh Pirates, whose current value is $62 million. Last year the team was valued at $70 million. What percent of decline did the Pirate organization experience over the past year?

____

8. In the NBA the New York Knicks franchise is considered the most valuable with a current value of $205 million, up 18% over last year. What was the team's value last year? _______________

9. The Big Apple (New York City) has many professional sports teams. If the Yankees are worth $209 million, the Knicks worth $205 million, the Mets worth 131 million, the Giants football team worth $183 million, the Rangers hockey team worth $118 million, and the New York Islanders worth $60 million, what is the total value of the professional teams in New York City? _______________

10. If the Golden State Warriors franchise grew from $93 million to $114 million and the Phoenix Suns franchise grew from $156 million to $191 million, which franchise showed the greatest percentage of growth during the past year?

____

11. The Chicago Bulls franchise showed moderate growth by going from a value of $166 million to $178 million during the current year. What was the rate of growth in the Chicago Bulls franchise over the past year? _______________

Name ____________________

Take Me Out to the Crowd

Have you ever thought about how much money is made during a really big-time football game? And what about the biggest football game of them all—the Super Bowl! It is the most sought-after ticket in sports. It also represents the most sought-after television commercial air time of any sporting event. How really big all these numbers are can only be understood by analyzing the buying habits of the fans.

A recent Super Bowl was played in Sun Devil Stadium in Tempe, Arizona. The seating capacity of the stadium is 73,521. Keep this statistic in mind as you work through the problems below.

1. If the "average" price of a ticket to see the "big game" is $215, what is the total amount of money realized from the ticket sales? ______________

2. It is almost an American tradition to eat a "stadium dog" when going to the ballpark. There's something about "being there" that makes those hot dogs taste so good that they are almost a part of the game itself. If there are .925 consumed per person . . . and they cost $3.00 per person, how much money is realized off the sale of hot dogs alone at the Super Bowl? ______________

3. Soft drink sales soar at such an important game, especially one played in the desert! If the total number of soft drinks consumed is 1.13 drinks per person, and the average drink consumed measures 26 ounces, how many gallons of soft drinks were sold during this Super Bowl? ______________

4. Program sales become another major source of revenue at such an important game. After all, who could afford to spend this kind of money to buy a ticket and travel there to the game without coming away without a commemorative souvenir program. If the programs sold for $5.00 each, and 77% of the people in the stands bought programs, how much money was realized from the sale of programs alone? _______________

5. Parking represents yet another source of revenue for any big-time sporting event. If parking for the Super Bowl averaged $6.00 per car, and there were 3.5 persons per car, how much money was collected from parking alone at this recent Super Bowl? (This would also account for bus traffic to and from the game.) _______________

6. Football fans are no different from other fans who attend major sporting events. They are in their seats for much of the game, but they also need to "stretch their legs" during the game, usually on more than one occasion. During that time, they may go to the concession stand, visit the restroom, or simply wander around a bit. If 93.2% of those in attendance were focused on the game itself at any given moment, how many fans could be found "wandering" at any given moment? (Round off to the nearest whole person.) _______________

7. T-shirts and other wearing apparel bearing the team and game logos represent yet another source of revenue at a game such as this. If 68.08% of the fans wound up with T-shirts, how many T-shirts were sold at this game? _______________

   If the T-shirts averaged $22 per shirt, how much money was collected from the sale of T-shirts? _______________

8. Fans come to the Super Bowl from all over the world. If the average distance traveled by fans to Tempe was 1,214 miles, what is the total number of miles all the spectators traveled to get to the Super Bowl? _______________

Name ____________________

Buy Me Some Peanuts and Crackerjacks!

Not so many years ago a box of Crackerjacks could be purchased for a quarter. Not so any more . . . especially at the old ballpark. Escalating costs, prompted by the extremely huge salaries paid to those who play the game, have forced team owners to look to any sources of revenue they can get!

One of those sources is the concession stand, where fans cool their thirst and whet their appetites by standing in lengthy lines paying outrageous prices for food that sometimes is less wonderful than it should be. But everyone does it, regardless, as the concession stand is all a part of "the old ball game."

Four fans were recently standing in line #124 and ordered the following from the concession stand. Look closely at the total price of each combination and then figure out what a single serving of each item costs.

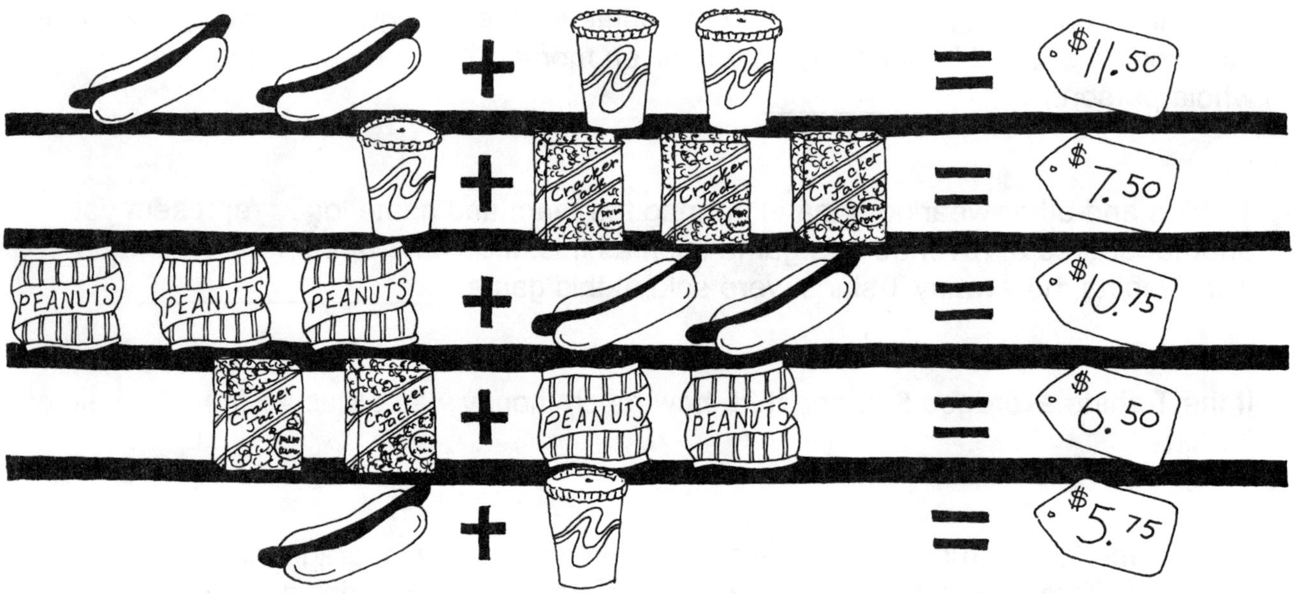

How much for a hot dog? __________ How much for a bag of peanuts? __________

How much for a box of Crackerjacks? __________

How much for a 46-ounce soft drink in a souvenir cup? __________

Name ______________________

$uper Commercials

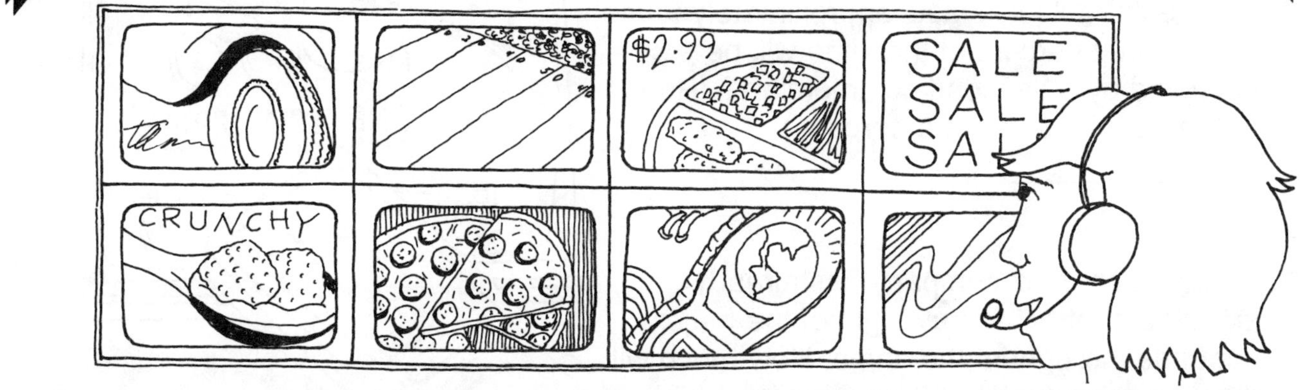

The high cost of advertising is staggering! While all TV ads involve considerable expense, the variable that most determines the cost of an ad is the number of people watching. The cost for a 30-second commercial during a local TV station's ten o'clock news costs far less than a commercial of the same length on national television during prime time viewing hours.

One of the big draws for spending money on advertising is sporting events. And there is none bigger than the Super Bowl! The cost of a 30-second spot ad during the "big game" two years ago cost $900,000! Last year that same spot cost $1.0 million. This year's price tag is $1.2 million for a 30-second spot ad. Consider this information in working the following "pricey" problems.

1. At a cost of $1.2 million for a 30-second commercial, how much money is each second of advertising costing the ad sponsor? ____________

2. Two years ago that same 30-second spot ad cost $900,000. What percent of increase does this year's cost represent over the cost two years ago? ____________

3. Last year's cost for the same 30-second commercial was $1.0 million. What percentage increase does this year's ad represent over an ad placed during last year's Super Bowl? ____________

4. There are 56 national advertising segments available for each Super Bowl. What will be the total revenue the network will receive for this year's Super Bowl?

____

5. How much time will be devoted to national television ads running 30-second commercials during this year's Super Bowl?

____

6. If the entire telecast lasts approximately four hours, what percent of that time will be occupied by 30-second commercials? (Round to the nearest tenth of a percent.)

____

7. A major soft drink company has already signed a contract to buy eight of the major spot ads throughout the game. How much will the company spend on commercials during this year's Super Bowl? _______________

8. A major fast-food hamburger franchise wants ten ad units. How much money will the company pay for these ads?

____

9. The cost of the television time for airing the ad is only a part of the total cost. Companies must spend a lot of time and money on the creation of the ad itself. The message to the viewing public must be both clever and appealing enough to get people to watch. Advertisers spend big bucks on the creative aspect of the ad and then must spend additional dollars on the actual staging of the project.

   To add to the high cost of advertising, companies often hire sports heroes to appear in their ads. If the fast-food hamburger franchise referred to in problem #8 spends $90,000 in production costs (including the advertising agency fee) plus an additional $150,000 to hire a big name tennis star and a basketball superstar to appear in the ad, how much will the cost of each commercial be if the company airs the same commercial six of its ten times? _______________

10. If you were responsible for developing an ad campaign for a series of Super Bowl ads, whom would you try to hire to appear in the ad to ensure that millions of Americans would watch? ____________________________

Name ______________________

Torch of Peace

The Olympic spirit began in Ancient Greece with a sacred truce among warring nations that would allow the athletes to compete against one another in peace. The sign of this truce was a torch that had been lit by the rays of the sun and signaled the beginning of the truce. The torch was carried by runners throughout the land.

The modern-day Olympic movement revived the tradition by using a torch to announce the opening of the 1936 Games in Berlin. The tradition has preceded every Olympiad since. During the 1996 Games in Atlanta, Georgia, the torch was brought from Olympia, Greece, to its journey throughout the United States to the eventual dramatic and emotional lighting of the Olympic cauldron in Atlanta on July 19. The long 15,280-mile journey, which passed through 42 of our states and Washington, D.C., began on April 27. Look closely at the map showing the route covered by the torch and answer the questions that follow.

1. The torch did not pass through Alaska and Hawaii for some obvious reasons. But six of the 48 contiguous United States did not see the torch either. Name those six states in the space below. To find the names of the states, look at a map showing state boundary lines.

   ______________________ ______________________

   ______________________ ______________________

   ______________________ ______________________

2. Why do you suppose these states were omitted? ______________________

   ____

3. How many days did it take for the torch to be relayed from Los Angeles to Atlanta?

   ____

4. Approximately how many miles was the torch advanced per day? ______________

5. The torch passed through the capital cities of many of our states. If you know the state capitals, mark through each and count the number of capitals through which the torch passed. (If you do not know, get a U.S. map and find out!) ________________

6. The rough terrain in certain areas of the United States combined with several heritage rich traditions in other areas resulted in methods other than runners carrying the torch some of the time. Included were railroad cars from a special Union Pacific train, a symbolic Pony Express segment, travel down the Arkansas River by Indians, and segments aboard a huge steamship on the Mississippi River. The remainder of the route was covered by the 10,000 runners, who ran no more than 1 kilometer (.62 miles). With this in mind, how many miles was the torch carried by means other than by runners?________________

7. If the torch were moved at a rate of approximately 14 miles per hour, approximately how many hours per day was the torch on the move? ________________

Torch Relay Route

Name ______________________

The Fastest Man on Earth

The current world record holder of the 100-meter dash is generally regarded as the world's fastest human being. There are several conditions (one of them is too strong a following wind) that place limits on accepting a recorded time as a world record. When a world record is recorded during the Olympic Games, it is a really big deal because the whole world is watching! Just how fast a human being can run such a short distance remains a mystery.

During the 1896 Olympic Games, Tom Burke of the United States won the gold medal with a time of 12.0 seconds. Four short years later Frank Jarvis of the United States recorded a winning time of 11.0 seconds! Since then the record has continued to fall as technology, better training techniques, advanced training diets, and other factors have combined to help the very best to become even better.

1. In 1968, Jim Hines of the United States established a world record in the 100 meters with a time of 9.95 seconds. How many meters was Hines traveling per second during this record-breaking effort? ______________

2. In 1988, Carl Lewis of the United States broke the existing world record by covering the 100-meter distance in a time of 9.92 seconds. How does his time compare with that of Jim Hines? (How much distance more per second was Lewis travelling than Hines?) ______________________________________

   ____

3. Before Americans began to adapt to the metric system, most track-and-field events in the United States were timed and measured under the United States conventional system, which included inches, feet, yards, and miles. Track meets in the United States typically included a 100-yard dash. When the metric system became the "accepted" universal measure to be used in track and field meets, people in the United States came to identify the 100-meter dash with a distance of 110 yards under the old American system. Though the two distances are comparable, which is longer? How much longer? Include in your answer the method of computation and line of thinking you used to arrive at your answer. ______________________

   ____

   ____

4. Reconsider the world record of 9.92 seconds for 100 meters held by Carl Lewis. How many feet per second was Lewis travelling during this historic sprint?

____

5. During the same time period, Gail Devers of the United States covered the same 100-meter distance in 10.82 seconds to become the fastest woman to run this distance. How many more feet per second was Carl Lewis running during his record run than Gail Devers? _______________

6. Convert the run of Carl Lewis to a miles-per-hour answer. How fast was he going?

____

7. The fastest 200-meter sprint in the Olympic Games belonged to Carl Lewis with a time of 19.80. The fastest 400-meter dash on record was that of Quincy Watts of the United States with a time of 43.50 seconds. Which runner was faster, Carl Lewis in the 100-meter dash, Carl Lewis in the 200-meter dash, or Quincy Watts in the 400-meter dash? _______________

8. How fast can you run 100 meters? Predict your time; then actually go out and run 100 meters and have your time recorded. How does your performance compare with those of Carl Lewis and Gail Devers? How many meters per second were you traveling during your run? _______________

9. How does your time compare to those of Tom Burke of the United States in 1896 with a time of 12.0 seconds and that of Betty Robison, also of the United States, who won the first women's 100-meter dash during the 1928 Olympics with a time of 12.2 seconds? _______________

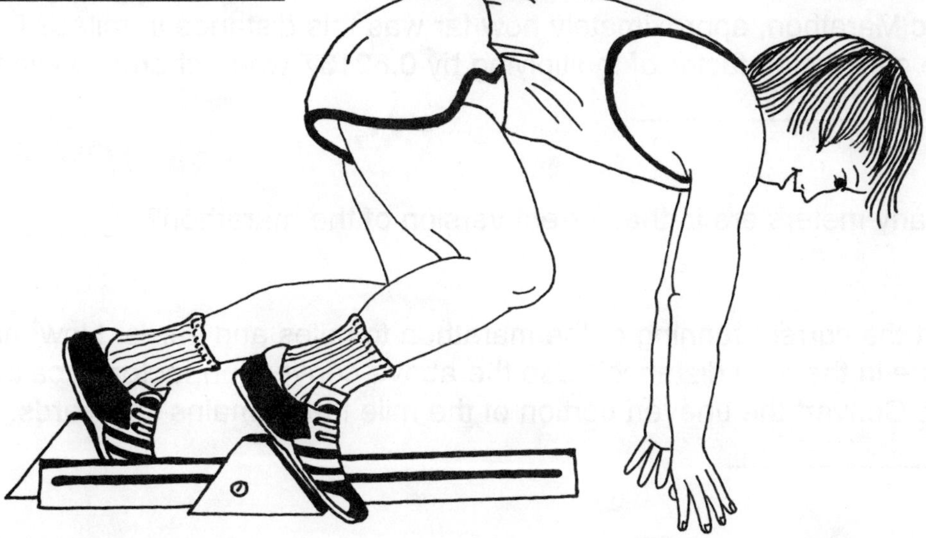

Name ____________________

Marathon Man

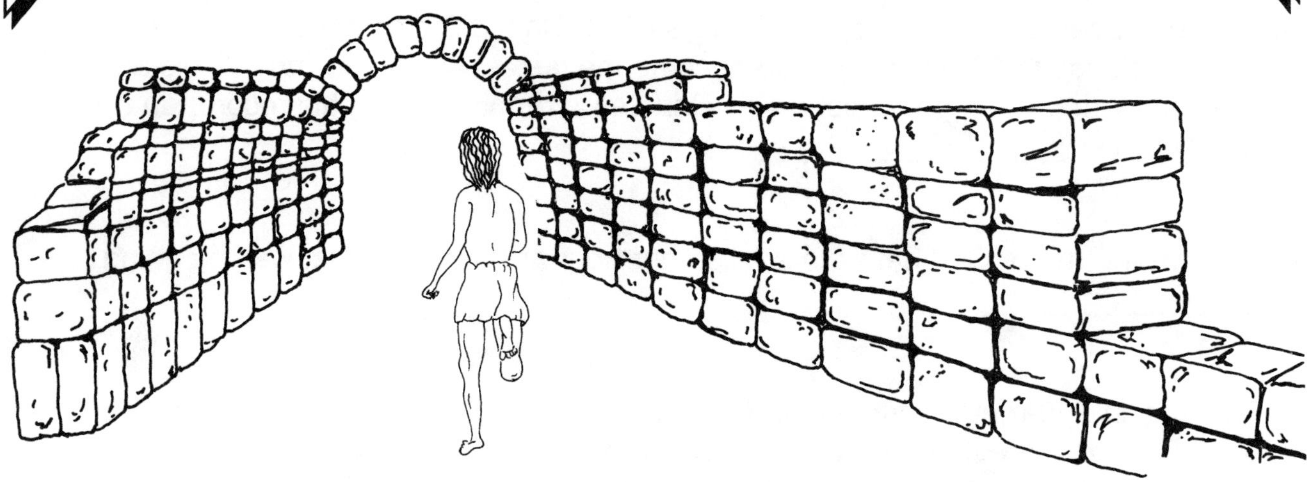

According to legend, in 490 B.C., a Greek soldier named Pheidippides ran from the site of the battle of Marathon to Athens, a distance of 36.2 kilometers. When he arrived in Athens, he was totally exhausted and announced the Greek victory over the Persians, then died. The modern Olympic Marathon commemorates that historic run. The marathon is regarded as one of the most prestigious of Olympic gold medals because of the strength and endurance required for such a race.

While the distance from Marathon to Athens is 36.2 kilometers, the modern version of the marathon was run at various distances until 1924, when the distance was established at 42.19 kilometers. That distance resulted from the British Olympic Committee's desire to have the race begin at Windsor and end at the royal box in London's White City Stadium.

1. If the distance from Marathon to Athens established the history behind the modern Olympic Marathon, approximately how far was this distance in miles? Be certain to use the conversion factor of multiplying by 0.62137 when changing kilometers to miles. ________________

2. How many meters are in the current version of the marathon? ________________

3. Convert the current running of the marathon to miles and yards. How many miles and yards are in the total distance? Use the above conversion factor to calculate your answer. Convert the uneven portion of the mile that remains into yards.

____

4. In 1972 Frank Shorter of the United States won the marathon with a time of two hours, 12 minutes, and 19.8 seconds. Rounded off to the nearest mile per hour, how fast was Frank Shorter running? ________________

5. The fastest marathon was run by Carlos Lopez of Portugal in 1984 with a record-setting time of two hours, nine minutes, and 21 seconds. How long did it take Lopez to run a mile at this pace? ________________

6. The current distance of the marathon is 26 miles, 385 yards. Based on this assumption, how many miles per hour was Lopez running during his record-setting pace? ________________

7. In 1984 Joan Benoit of the United States ran the marathon in two hours, 24 minutes, and 52 seconds. Calculate her speed in miles per hour and round off to the nearest tenth of a mile. ________________

8. The 1984 Olympic Games in Los Angeles provided new Olympic records for both women and men in the marathon. How much faster was the winning time for men than the winning time for women? ________________

9. Have you ever thought about how long it might take you to travel the distance of a marathon? Make a "guesstimate" in the space below and compare with other classmates. ________________

How long do you think it would take you to ride your bicycle 26 miles and 385 yards?

____

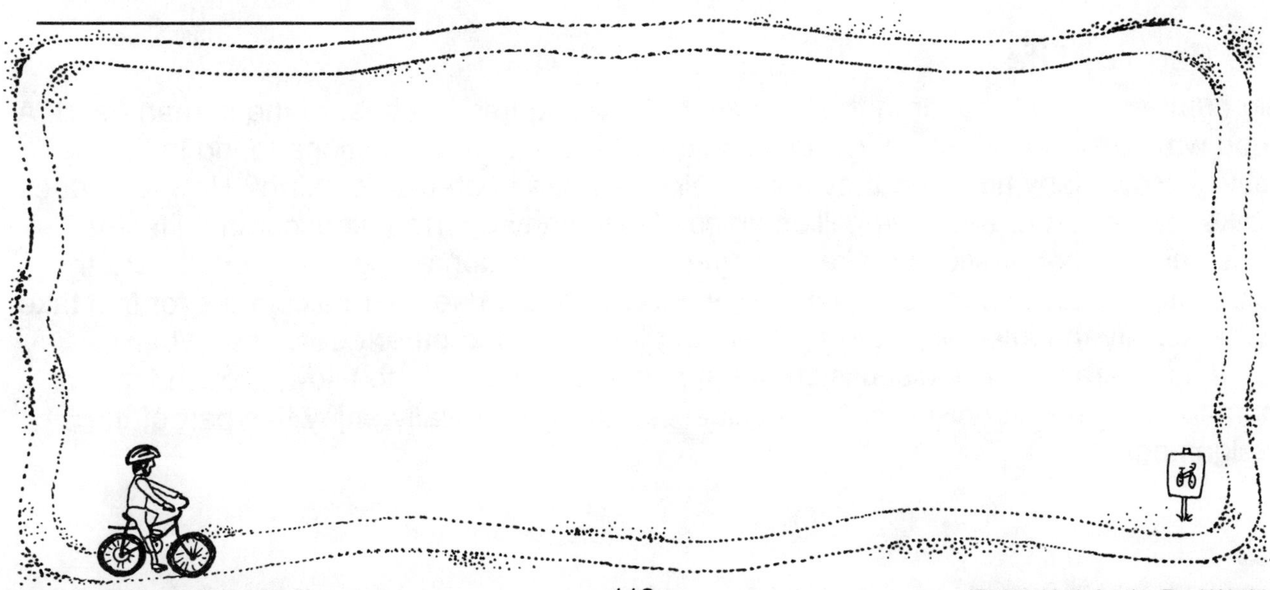

Teacher Notes

MYSELF AND OTHERS

The activities in this unit are all focused on facts, statistics, and numbers about students themselves and their classmates. Most middle school students find a great deal of interest in anything that involves using themselves as the center of the interest. Some of the activities involve their own personal preferences and choices and involve survey-type activities in which information is collected from home and contact with other students. As with other units and activities in this book, there is an abundance of "everyday math" that gives students reasons for its importance and direct association with the real world.

The Beat of Life

Big numbers again appear in this activity that investigates the beat of the human heart. A good way to start this activity is to ask students some of the questions found in the activity. How many times do they think their heart has beaten since birth? How long does it take for a heart to beat one million times? How many heartbeats occur in a lifetime? These and other questions make for some lively discussion and guess work before the assignment actually begins. If possible, students should use their calculators for this take-home activity that discovers the answers to the above and other questions. When students return for class discussion of the answers, be certain to allow enough time for them to compare and discuss the last question. Students really enjoy this part of the assignment.

Pound for Pound

This activity calls for analysis and study of data presented on a recent weight study of random Americans. To help with the organization of the data, it is suggested that students work in pairs. There is an amount of search-and-find within the data presented and four eyes are better than two! Even though students work in pairs, each has his/her own activity sheet to complete as some of the questions call for individual thinking.

Burning Calories

With the continual bombardment of information that emphasizes the value of physical exercise in maintaining good health, students are already well aware of the relationship between performing exercise and burning calories. The data presented in the chart and the questions that follow will help them to develop a more realistic idea of which exercises are best and learn about definable rewards for time spent doing various athletic activities.

About My Class

The activity is a good one to present near the beginning of the school year as it gives students a chance to tell a little about their own lives and preferences and to learn about the lives and families and preferences of other members in the class. It then becomes a class activity in which students pool their data to come up with some "whole class" numbers and statistics.

Battle of the Math Classes

Launch this assignment by asking students how long of a line they think could be formed if all students in the class were placed in a long row on the floor, head-to-toe-to-head-to-toe until all students were included. After accepting a number of guesses, tell them that is one of the things they are about to find out with this assignment. Divide students into pairs and provide them with calculators if possible. Tell them that they are free to find out the total height of the students in each class any way they may desire. The culmination of the activity is to measure your own class to find out how it "stacks up."

Tube Time

Most of us are very aware that we spend too much time watching television. With the addition of subscription channels, movie channels, sports channels, home shopping channels, and all the other channels that cater to specific interests and desires, there's really something for everyone on television, or so it seems. Students become involved with their own families in logging their viewing habits for an entire week. Emphasize that this isn't to be a burdensome task for them or their parents. They simply jot down the times each day when they are watching television. That's it. Students then total the time and perform the math that will analyze the data. Allow students to share their own personal evaluations of their televiewing habits with other members of the class.

A Slice of Life

Since students found out in Tube Time just the percentage of their week they spend watching television, a good way to initiate this activity is by posing other questions that relate to how students spend their time. How much time do they spend sleeping? Eating? Playing? Doing family chores? Finding the answers to these questions is the focus of this activity. Students are asked to keep another log, this one their own personal accounting for a 24-hour period of their life. Emphasize that it is unimportant to jot down EVERYTHING. Once the logs are completed, have each student organize the data and answer the questions on the third page of the activity. The bar graph that becomes a circle graph is a neat way to avoid all that complicated math!

Higher Learning—Higher Costs

With the cost of going to college escalating at an alarming rate, students need to give some thought about how they plan to get there. While new studies on college costs are released almost as often as the costs themselves rise, the numbers used in this activity were the most current at the time of publication. Explain to students the need for factoring in an inflation factor and also a rate of return on money. Even though students may not know how those factors are derived, they can understand the concept of how they work. Go through the example with them that is provided at the top of the second page. It will help them to understand what they are supposed to do on the second page of the assignment. Allow them to share their own thoughts on college (if they like) with other members of the class.

Name ______________________________

The Beat of Life

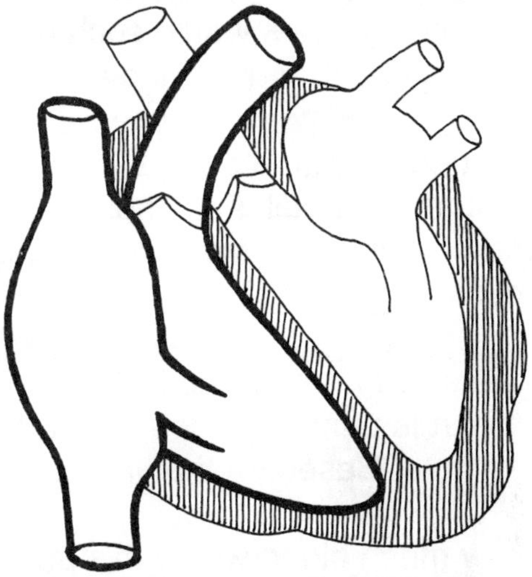

The human heart is an amazing four-chambered muscle that beats without stopping for an average of more than 70 years, continuously pumping millions of quarts of blood through our bodies. At rest, the heart pumps approximately two ounces of blood per beat and around five quarts per minute, compared to approximately 4-7 ounces per beat and 21-32 quarts per minute during vigorous exercise. The adult human heart is about the size of a fist and weighs about nine ounces.

To find out how really hard your heart works for you every hour of every day of your life, work the following problems.

1. As we all know, the heart goes on and on and on, beating tirelessly for many many years. If a normal heart beats around 70 beats per minute, which of the following would be your best guess on how long it would take for it to beat 1,000,000 times? a day . . . two days . . . a week . . . ten days . . . two weeks . . . a month . . . a year . . . ten years? Do not actually calculate at this time. Just make an educated guess and jot your answer down here. ______________________________

2. If the normal heart beats around 70 beats per minute,

   a. how many times would it beat in an hour? ____________

   b. how many times would it beat in an entire 24-hour day? ____________

   c. how many times would this heart at rest beat in a week? ____________

   d. how many times would it beat in an entire year? ____________

   e. how many heartbeats could we expect of this heart in the entire lifetime of a person whose heart averaged 70 beats per minute for 75 years? ____________

3. Go back to problem #1. Which of the answers seems the most reasonable?

   ____

4. Strenuous exercise can increase the heartbeat to up to 200 times per minute. When the person (whose heart normally beats 70 times per minute) stops exercising, the heartbeat usually lowers to an average of approximately 117 times per minute. Given that information, answer the following:

   a. How many times would such a heart beat in the approximately 15 minutes it would take the heart to return to a normal beat? _______________

   b. If a person with a normal rate of 70 beats per minute exercises strenuously for 30 minutes and builds the heart rate up to speed of 180 times per minute, how many more heartbeats will occur during that time of exercise than when the person is at rest? _______________

5. If a person's normal rate is 70 beats per minute, the heart is pumping about 5 quarts of blood per minute. Based on that information . . .

   a. How much blood would this person's heart pump in a 24-hour day? _______________

   b. How much blood would such a heart pump in a year? _______________

   c. Based on this same rate, how much blood would a heart pump through a body that lasts 78 years, 6 months, and 3 days?

   ____

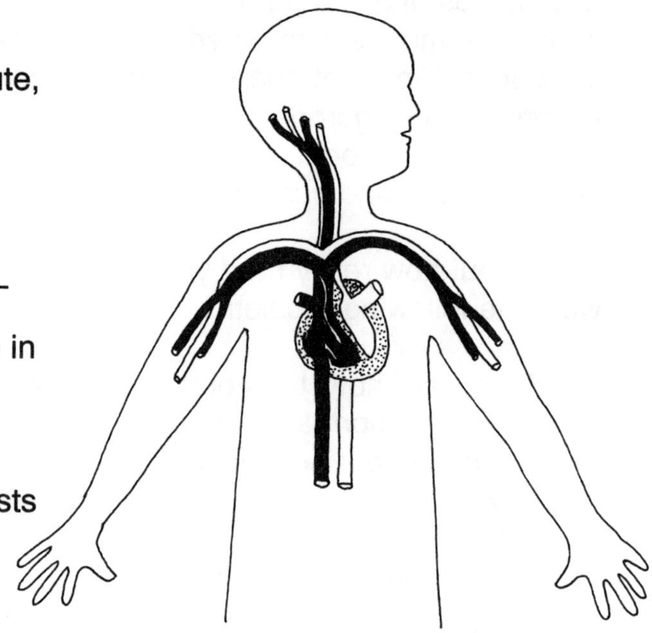

6. Strenuous exercise can induce the heart to pump up to 30 quarts of blood per minute. Based on this statistic, how much blood would a marathon champion's heart be expected to pump during a race in which a time of 2 hours, 15 minutes, and 30 seconds was recorded? _______________

7. Choose a partner for this part of the activity. You will need a watch with a second hand or one that shows a digital reading for seconds. Take turns with your partner in recording each other's pulse for 30 seconds. To do this, place your forefinger and middle finger together on the underside of your partner's wrist (where you can feel the pulsating beat). Count the number of times his or her heart beats in 30 seconds; then multiply by two to get the number of beats per minute. Now calculate the following data on your own heartbeat (based on the findings of your partner).

   a. How many times would your heart beat in an hour? _______________

   b. How many times would your heart beat in a year? _______________

Name ______________________________

Pound for Pound

Americans are gaining weight! During a study conducted during a recent year, it was discovered that nearly 31 percent of adult men and 35 percent of adult women were overweight! That figure represents a 30 percent increase in overweight people since 1980! Being overweight can lead to a number of serious health problems that include diabetes, cardiovascular disease, gallbladder problems, and even certain kinds of cancer. Studies have concluded that the main reasons for so many Americans being overweight come down to two problems: we eat too much and we don't exercise enough!

Another recent survey reported that only about one in three adult Americans engages in strenuous exercise at least three days per week. Worse yet, many do not exercise at all!

The first step in solving the problem of being overweight is to determine just how much overweight an individual really is. The next step involves coming up with a solution that will include realistic goals while at the same time helping the individual to return to a more normal weight. Below is a list of the heights, weights, and ages of 40 people.

Height	Age	Weight	Height	Age	Weight
5'4"	26	162	5'6"	39	174
5'8"	21	161	5'8"	62	174
6'0"	24	181	6'2"	56	210
5'2"	34	141	6'2"	49	228
5'6"	50	178	5'9"	61	206
5'11"	66	209	6'4"	29	195
5'8"	42	178	5'4"	53	149
6'2"	27	204	6'0"	29	210
5'6"	33	198	6'1"	50	219
5'7"	39	188	5'10"	29	180
5'1"	20	142	6'5"	20	224
6'2"	70	193	5'5"	51	170
5'8"	29	179	6'1"	47	179
5'2"	19	144	6'1"	64	204

Height	Age	Weight	Height	Age	Weight
6'0"	28	194	5'9"	61	191
5'4"	24	150	5'6"	44	169
5'11"	48	186	6'0"	59	206
6'2"	54	214	5'8"	51	160
6'6"	33	220	6'2"	50	211
5'8"	27	168	5'10"	58	177

The following chart was released to help overweight people assess their own weight problems. The chart is based on height and weight, but it also includes an age factor that breaks down into two age groups. Note also that there is a fairly broad range of weight in the "ideal" category at each age and weight level.

Height	Weight in Pounds	
	19-34 years	35 years and older
5'0"	97-128	108-138
5'1"	101-132	111-143
5'2"	104-137	115-148
5'3"	107-141	119-152
5'4"	111-146	122-157
5'5"	114-150	126-162
5'6"	118-155	130-167
5'7"	121-160	134-172
5'8"	125-164	130-167
5'9"	129-169	142-183
5'10"	132-174	146-188
5'11"	136-179	151-194
6'0"	140-184	155-199
6'1"	144-189	159-205
6'2"	148-195	164-210
6'3"	152-200	168-216
6'4"	156-205	171-222
6'5"	160-211	177-228
6'6"	164-216	182-234

1. How many of the people in this study are considered overweight according to the chart above? _______________

2. What percentage of the people in this study are overweight? _______________

3. How does this survey of Americans compare with the national averages on overweight people? _______________

4. Go back through the chart and list below the three most overweight people in the study.

   ____

   ____

   ____

5. Identify the two people who could be considered to have the "most ideal" weight.

   ____

   ____

6. Why do you suppose the ideal weight for people between the ages of 19-34 is somewhat less that the ideal weight for people 35 years of age and older?

   ____

   ____

7. What conclusions can you draw from this data?

   ____

   ____

   ____

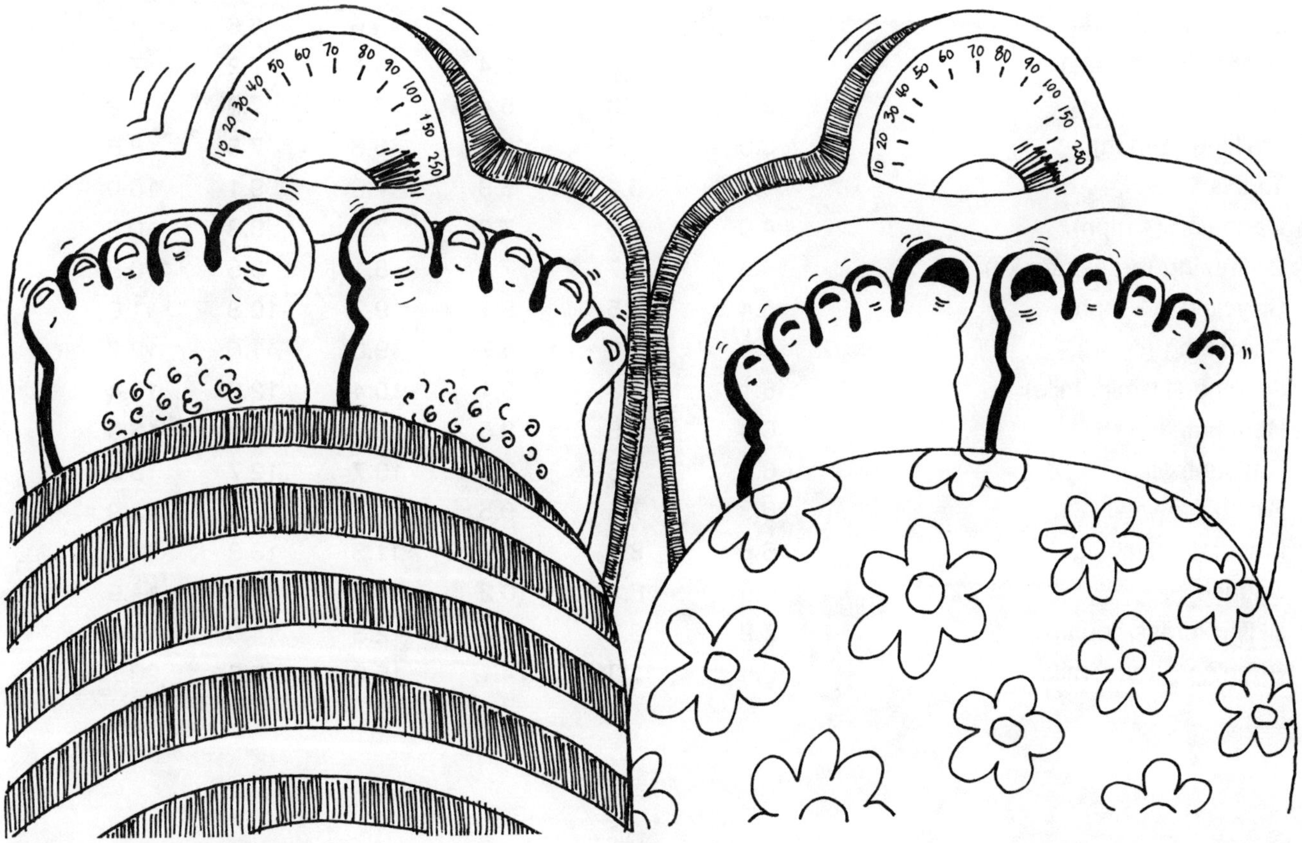

Burning Calories

While millions of Americans remain overweight because they eat too much and do not get enough exercise, there are millions of other Americans who have become very health conscious and are aware of the many health benefits associated with eating properly and exercising on a regular basis. A good diet and regular exercise are great companions to have in building strong bones and warding off disease, and they also contribute to a good mental outlook on life itself.

Below is a chart showing the number of calories burned during each minute engaged in a number of athletic activities. Note also that various body weights are included on the chart. It stands to reason that the more a person weighs, the more calories are actually burned. It also becomes obvious that the more strenuous activities burn calories more quickly. Look closely at the data presented in the chart and answer the questions that follow.

Activity	100 lbs.	120 lbs.	150 lbs.	170 lbs.	200 lbs.	220 lbs.
Volleyball (moderate)	2.3	2.7	3.4	3.9	4.6	5.0
Walking (3 mph)	2.7	3.2	4.0	4.6	5.4	5.9
Table tennis	2.7	3.2	4.0	4.6	5.4	5.9
Bicycling (5.5 mph)	3.1	3.8	4.7	5.3	6.3	6.9
Calisthenics	3.3	3.9	4.9	5.6	6.6	7.2
Skating (moderate)	3.6	4.3	5.4	6.1	7.2`	7.9
Golf	3.6	4.3	5.4	6.1	7.2	7.9
Walking (4 mph)	3.9	4.6	5.8	6.6	7.8	8.5
Tennis	4.5	5.4	6.8	7.6	9.1	10.0
Canoeing (4 mph)	4.6	5.6	7.0	7.9	9.3	10.2
Swimming (breast stroke)	4.8	5.7	7.2	8.1	9.6	10.5
Bicycling (10 mph)	5.4	6.5	8.1	9.2	10.8	11.9
Swimming (crawl)	5.8	6.9	8.7	9.8	11.6	12.7
Jogging (11-min. mile)	6.1	7.3	9.1	10.4	12.2	13.4
Handball	6.3	7.6	9.5	10.7	12.7	13.9
Racquetball	6.3	7.6	9.5	10.7	12.7	13.9
Skiing (downhill)	6.3	7.6	9.5	10.7	12.7	13.9
Mountain climbing	6.6	8.0	10.0	11.3	13.3	14.6
Squash	6.8	8.1	10.2	11.5	13.6	14.9
Skiing (cross-country)	7.2	8.7	10.8	12.3	14.5	15.9
Running (8-min. mile)	9.4	11.3	14.1	16.0	18.8	20.7

1. Which do you think is the more strenuous exercise, playing golf or walking at a rate of 3 miles per hour (20-minute miles)? ____________________

2. Which exercise burns the most calories, jogging at a rate of 11 minutes per mile or riding a bicycle at a speed of 10 miles per hour? ____________________

3. Is playing racquetball or table tennis better for burning calories? ________________

4. Why do you think the people who prepared this chart ranked handball and racquetball as "equal" exercises throughout the chart? ______________________________

____

5. Which is more beneficial exercise, swimming the breast stroke or swimming the crawl stroke? ____________________

6. Which exercise in the chart would be considered the most strenuous because of the calories burned? ____________________

7. Which exercise in the chart would be considered the easiest to do because it burns the fewest calories per minute?

____

8. How many calories would a 100-pound person burn during an hour of table tennis? ____________________

9. How many calories would a 150-pound person burn while paddling a canoe for 30 minutes? ____________________

10. How many calories does a 120-pound person burn while swimming the breast stroke for 40 minutes? ____________________

11. How many calories does a 220-pound man burn while engaged in a mountain climb that lasts three hours and 45 minutes? ____________________

12. How many more calories per hour does a 170-pound downhill skier burn than a 120-pound person engaged in the same exercise? ____________________

13. How many more calories does a 200-pound person burn playing two hours of racquetball than his 150-pound opponent? ____________________

14. Which exercise burns more calories for a 150-pound person, 30 minutes of squash or an hour of volleyball? ____________________

15. Which exercise burns more calories for a 200-pound person, one hour and 15 minutes of tennis or 45 minutes of handball? ____________________

16. How many more calories does a 120-pound bicycle rider burn during a two-hour ride at a speed of 5.5 miles per hour than a one-hour ride at 10 miles per hour?

____

17. How many more calories does a 170-pound runner burn when running five miles at a speed of 8 minutes per mile than the same jogger running the same distance at a speed of 11 minutes per mile? ____________________

18. How many more calories does a 150-pound downhill skier burn during two hours on the slopes than a canoer of the same weight? ____________________

19. Who burns more calories—a 220-pound table tennis player who plays for 45 minutes or a 100-pound jogger running at a pace of 11 minutes per mile for the same length of time? ____________________

20. How many more calories does a 200-pound golfer burn during a four-hour round of golf than a 170-pound person on a canoe ride that lasts three hours and 30 minutes?

____

21. In which of these athletic activities do you like to participate? List your favorites in the space below. ____________________

____

22. Choose your two favorites and make up your own math problem that will show a comparison of the calories burned between the two activities. ____________________

____

Name ______________________________

About My Class

Complete the survey below by collecting and reporting all the needed information. From this data, you will then participate in developing a composite that reflects the total combined data provided by you and your classmates.

1. How many people live in your house? Be certain to count yourself! ______________
2. What is your favorite color? ______________
3. What is your favorite sport? ______________
4. What is your favorite TV program? ______________
5. How many televisions are there in your home? ______________
6. How many cars does your family have? ______________
7. How many pets live at your house? ______________
8. What is your favorite subject in school? ______________
9. Is there a computer in your home? ______________
10. What is your favorite holiday? ______________

From the data provided by all members of your class, answer the following:

1. What is the sum total of all the people living in all of the homes occupied by all of the members of your class? ______________

2. Find the mean number of people living in each house. Round off to the nearest hundredth. ______________

3. What is the average number of televisions per house? ______________

4. Which holiday was the "class favorite"? ______________

5. What percentage of homes have computers? ______________

6. How many different "favorite sports" do your classmates have? ______________ Which was the most popular? ______________

7. Is there a difference between your favorite sport to watch and your favorite sport to play? ______________

8. How many students in your class have at least one pet? ____________ How many have more than one? ____________ Are dogs or cats more popular pets with the students in your class? _________________

9. Which subject ranked highest among the members of your class? ______________

   How many students chose that subject? ______________

10. Was there a clear-cut favorite television show among the members of your class?

    ____

    If so, what was it? ______________

Choose from all the data that has been collected and portray the information gathered from the answers to ONE QUESTION by creating a bar graph. Be certain to label your graph carefully to make it easy for others to read.

Name ______________________

Battle of the Math Classes

Miss Dillahunte assigned her math class the task of finding out the average height of the students in her class. Tami was sharing "her day at school" with her family at dinner and recapped the assignment, telling everyone at the table she thought it "sounded like fun."

Tami's twin brother Tony said that he "bet the kids in Mrs. Little's math class were taller than those in Miss Dillahunte's class." Tami said she really didn't think he was right, and that perhaps Mrs. Little's class could do the same as Miss Dillahunte's and the two could then compare. The next day Tony proposed the plan to Mrs. Little and she thought it was a great idea.

Each class set about making their own arrangements to measure the height of each student. To make the measurements consistent, it was decided that all students would remove their shoes when they were being measured. Following are the results of the measurements of the two classrooms:

Miss Dillahunte		Mrs. Little	
Maria	5'½"	Tommie	5'2"
Paul	5'3¼"	Tony	5'½"
Melissa	4'11½"	Tracie	5'½"
Wendy	5'2¾"	Randy	5'3½"
Phillip	5'2½"	Robyn	4'11"
Stacie	5'¾"	Kevin	5'4"
Raol	5'2¼"	Joseph	5'2"
Dawn	5'3"	Janet	5'2¾"
Jeff	5'5"	Andrew	5'3¾"
Michael	5'1"	Linda	5'¼"
Mark	4'11"	Peter	5'4"
Shawn	5'0"	Brandi	5'2½"
Tyler	5'6½"	Nancey	5'3½"
Betsy	5'2½"	Frankie	5'2¼"
Terry	5'2¾	Frances	5'0"
Marvin	5'½"	Jerry	5'6½"
William	5'4½"	Andy	5'4½"
Joey	5'2½"	Judy	4'10"
Zoey	4'10½"	Janie	5'3¼"
Laurie	5'4"	Sandi	5'3¾"
Jacob	5'0"	Missy	5'2"
Mandy	5'3"	Annie	5'3¼"
Ron	5'6"	Billy	5'6¼"
Phyllis	5'2¾"	Derrick	5'5'

1. How "tall" is Miss Dillahunte's class of students? What is the sum of all her students' heights? Express your answer in feet and inches. _______________

2. How "tall" is Mrs. Little's class—the sum of the heights of all her students? Again, express your answer in feet and inches. _______________

3. Which class is taller? How much taller? _______________ _______________

4. If the total heights of students in both classes were added together, what would the distance be? _______________

5. What can you think of in the real world that would be about the same distance?

____

6. What is the average height of a student in Miss Dillahunte's class? Express your answer in feet and inches, rounding off to the nearest one-quarter inch.

____

7. What is the average height of a student in Mrs. Little's class? Express your answer in feet and inches, rounding off to the nearest one-quarter inch. _______________

8. Conduct your own research and create your own data by performing a similar experiment within your own class.

Name ________________________________

Tube Time

Most of us spend entirely too much time in front of the television. There certainly *are* many programs and shows well worth watching that educate and entertain us. But there is a great deal of programming on today's television that really doesn't merit our spending our time. During the early 1990s American televiewing habits actually went into a slight decline as people began to spend more of their time on home computers.

But a recent survey indicates that television is on its way back! With the cable and satellite business booming, Americans are once again turned on to TV. More subscription channels, more movie channels, more home video shopping channels, and basically more channels period have us glued to the set more hours than we should be!

Have you ever asked yourself how long you spend during an average day watching programmed television? How about your friends? How about the rest of the members of your family? The whole purpose of this assignment is to find answers to these and other questions. Here's how it works:

Look at the TV Log that accompanies this assignment. You are going to keep this log for an entire week, recording the times when you are watching television. You will also get a log for each member of your family. Ask them to do the same thing as a favor to you in helping you to complete this assignment. Here are the ground rules:

A. Rented videos and video arcade games don't count. This assignment is only concerned with keeping track of the time you spend actually watching programmed television.

B. To simplify the assignment, you and your family need only to record time in 15-minute segments. If your mom just turns on the TV to the weather channel to see what kind of a day it's going to be, that doesn't count either. Record only 15-minute segments or more.

C. Record the beginning and ending time of each viewing segment. For example, if you get home from school and watch a program from 4:30-5:00 P.M., you should record it as such, then indicate 30 minutes as the total time. Round off all viewing times to the nearest quarter hour.

D. Start the assignment on a Sunday, and keep the log (and encourage your family to do the same) for an entire week that concludes with the end of your day Saturday. Make no mention of the kind of programs you or other members of your family watch. That is another survey entirely. You are only recording actual viewing time.

E. Once you have collected the TV Viewing Logs from all family members, your next task is to total all of their viewing time during the past week plus yours. Take this data to school with you the following Monday to be entered into a survey of your entire class.

F. Answer these questions below that summarize the results of your own family's television viewing.

1. How much total time did you yourself spend in front of the television during this past week? ____________

2. On which day did you spend the most time watching TV? ____________

3. In the space below, jot down the name of each family member including yourself and record the total time of each family member.

   ____

   ____

   ____

4. Which family member spent the most time watching television during the past week?

   ____

5. Calculate the number of hours in an entire week and record that number here.

   ____

6. What percentage of your past week was spent in front of the TV set?

   ____

7. Write your own summary reaction to the data you have collected. Do you think your viewing habits are about average, above average, or below average? What about the other members of your family? Record your thoughts on the reverse side of this page.

   ____

   ____

______'s

T.V. Viewing Log

	times watched	daily total
Sunday		
Monday		
Tuesday		
Wednesday		
Thursday		
Friday		
Saturday		
	Weekly total	

Name ____________________________

A Slice of Life

Some middle school students have the feeling that everyday living can become pretty "ho-hum." Yesterday is often like today, which is many times just like tomorrow. What they really should realize is that we all have a limited number of days in our lives and none of them should be wasted. The whole purpose of this activity is to get you to take a few moments and analyze the way you spend most of your time to make certain that you are spending it the way you want it to be spent.

Have you ever really thought about this before? How much time do you spend sleeping? How much time do you spend eating? How much of your day is taken up by school? How much time do you have each day to do what you would really like to do? These and other questions are what this assignment is all about.

There are 24 hours in each day of every day of your life. You obviously have certain commitments over which you have no control that occupy part of your day. But you are also left with a lot of free time where the choice is really yours. How you spend that time is up to you! For example, if you slept 12 hours each day, you would be spending half of your life sleeping! You obviously don't do that . . . but how much time each day do you really spend sleeping?

It's a good experience to actually monitor the way you spend your time during a typical 24-hour period of time. You and the other members of your class will be basing your data collecting on actions and events that really happen.

Use the chart on the next page to record how you spend your time over the next 24 hours. Include a running log from the time you get up tomorrow morning until the time you go to bed. Then your 24-hour day will end with the time you get up the next day. Be certain to include anything "out of the ordinary" that happens to you that might affect your total data. To make your job a little easier, round off all activities to the nearest quarter-hour. There will be some "quick" events that you will obviously not need to record.

DAY OF THE WEEK CHOSEN: ______________________________

RISE AND SHINE!

TIME OF DAY	ACTIVITY

Answer the questions below based on the information you collected over the last 24-hour period when you were keeping track of how you spent your time.

1. How much time did you spend sleeping? _______________

2. How much time did you spend eating? _______________

3. How much of your time did you spend at school? _______________

4. How much time did you have to really do what you wanted do? _______________

5. Did you watch any television? If so, how did that "TV time" rank with your viewing habits which you recorded in the "Tube Time" activity?

   ____

   ____

6. Your final task in this assignment is to create a circle graph that will give you a graphic picture of how you spent each "slice of life" during the day in which you recorded your activities. To make the circle graph, you need first to make a bar graph. Total all of your activities and summarize into quarter-hour segments below. Then mark off each segment on the bar graph. Label each bar and color it. Use a different color for each bar. Then very precisely cut out each bar and tape them together. Finally, turn the taped-together bars into a circle that will provide the definition for your circle graph. Draw a circle on the reverse side of this paper that is similar to the circle created by your "connected bars." The final step is to simply mark off arcs on the circle that correspond to the ends of each bar. Label each portion of the circle to give you and others a better idea of how you spend your time.

Activity

Time Spent

Name ______________________

Higher Learning—Higher Costs

The cost of going to college is on the rise. Wow! Is it ever on the rise! Total costs for attending private colleges and universities have grown 95 percent since 1984. While not quite as bad, costs for attending public colleges and universities have risen 82 percent during the same time period. These trends have raised serious questions about the future affordability of higher education for many people.

Costs have skyrocketed at a time when a college degree has become almost a necessity for young people striving to succeed. The burden is being shared by middle class parents saddled with huge debts and the students themselves who go deeply into debt with student loans.

A recent survey announced the average cost of a single year at a private college or university was $17,631! A year at Princeton University costs $28,325, and a year at Duke University costs $27,799! Many of the more prestigious institutions of higher learning throughout the United States have similar costs. The average cost of attending a public four-year college or university is only $6,823, but that figure, too, has risen dramatically over the past few years. There are over 8.8 million students enrolled in the more than 2,200 four-year colleges and universities throughout the United States.

Many families begin planning the college futures of their children almost from the moment of birth. But those who make such plans must factor in a rate of inflation that reduces the value of current dollars.

Also to be figured is an annual expected return on the money that is saved. For example, parents who start saving for their son's education beginning on his second birthday, look to the following formula to tell them how much money they should be saving.

Their son will begin college at the age of 18 years. That leaves 16 years in which to save enough money. If the annual inflation rate is assumed to be 5%, then the factor to use is 2.18. That factor is multiplied by the current cost of attending a four-year public university ($6,823) to yield an expected per year cost of $14,874 sixteen years from now. Multiply that number by 4 (the number of years in college), and the total amount needed becomes $59,496. If the expected rate of return is 8% annually, then a factor of 32.75 will show how much the money will grow. Divide the total need ($59,496) by the factor (32.75), and the parents realize they need to save approximately $1,816.67 per year, not an unreasonable figure at all.

Use the same thinking to work the following problems that include some people who did not plan as well as the parents above.

1. Assume that parents of a six-year old want to start a savings program that will allow their daughter to enroll in a public college or university at the age of 18. If the inflation rate they use is 5%, the factor becomes 1.80. If they use a rate of return on their investment of 8%, the factor to use is 20.50. How much money should they be saving each year if they want a plan that will allow their daughter to have enough money when she finishes high school? _______________

2. How much money will be needed by parents of an eight-year old who want to send their child to a private college if they figure an annual inflation rate of 5% (and the factor is 1.62)? _______________

3. Parents of a 12-year old want to start saving for their child's college education at a private college. Figuring a rate of inflation of 5%, the factor is 1.34. They are assuming an 8% return on their money, which gives them a factor of 7.82. How much money will they need to save each year to have their child's college education paid for upon enrollment? _______________

4. Have you ever thought about your own plans for college? Where do you want to go? Do you know what kind of college you want to attend? Do you know what it currently costs per year, including tuition and room and board? Do you or your parents have any kind of savings plan for your college education?

____

____

____

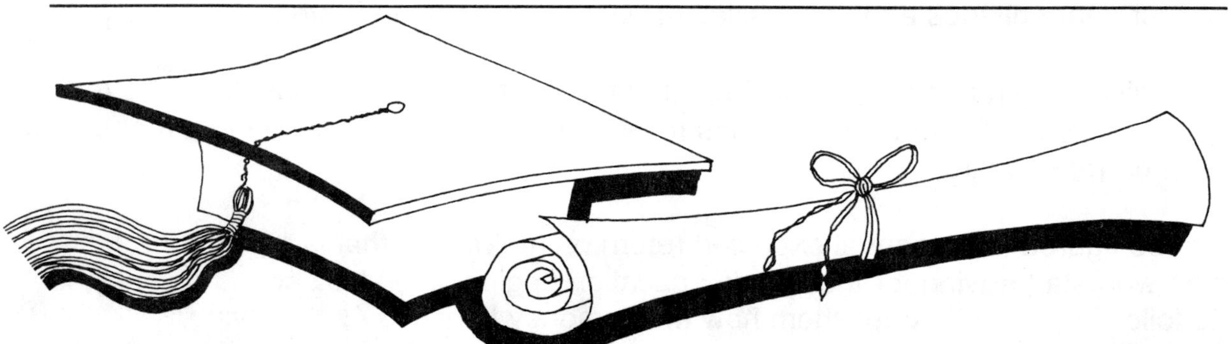

MATH, ET CETERA

This last unit contains a smorgasbord of math activities that offer additional variety to the already included units. Showing the relevance of math to the real world continues to be the focus and purpose of all ideas that are presented.

Our Tallest Skyscrapers

Students compare the heights of the ten tallest buildings in the United States. After rank-ordering them in descending height in the first part of the activity, they complete problems that relate to specific comparisons. Launch the activity by asking students if they know the name of the tallest building in the world (Sears Tower) and where it is located (Chicago). As a follow up, ask them to research the tallest free-standing structure in the world (CN Tower in Toronto, Canada, stands 1,821 feet high.)

Mother Knows Best

This activity presents students with a problem in finding the weight of each child without knowing the individual weight of any child. What they are provided is information of the combined weights of two children in various combinations. Strategies may vary, but most students will probably see the value in #2 if they find no other route to the solution.

Happy Birthday, Andy

This activity also asks students to find the individual ages of Andy, his mom, and his dad without knowing the age of any single individual. Students use the notebook approach of supplying the information they do know to help them in arriving at what it is they are supposed to find out.

Tickets, Please

Here is another problem in which students will probably use the notebook strategy. Based on the information provided, students should be able to calculate the various ticket combination prices.

Shades of Summer

There are several strategies that can be used, but comparing quantities by using the exchange method will probably be used by most students.

It's About Time

Before assigning this activity, give each student a copy of the time zones map that shows the various time zones found in the United States. Be certain they are familiar with how to use the map by offering a few simple problems in class as examples they can follow. The problems on the activity sheet vary in difficulty with the more difficult problems near the end of the assignment. Be certain students complete question #20, in which they are asked to create their own time zone problem and exchange with a friend.

What Are My Chances?

The entire assignment focuses on probability. Much of the assignment calls for students to work in groups. Provide each group with a pair of dice. They can use their own coin to satisfy question #1. Questions #2 and #3 relate to a deck of cards but none is required. Neither is a spinner required to answer question #5. Be certain students understand how to express the likelihood of an event happening by writing it down as a fraction reduced to its lowest terms. An extension of the activity might be to ask them to calculate the odds of winning the lottery given a stated number of numerical choices.

Name ____________________

Our Tallest Skyscrapers

Below is a list of the ten tallest skyscrapers in the United States. Take a close look at how they stack up. Rank order them from the tallest (#1) to the tenth tallest (#10).

_____	Empire State Building, New York	1,250 feet
_____	Nations Bank, Atlanta	1,027 feet
_____	Chrysler Building, New York	1,046 feet
_____	Sears Tower, Chicago	1,454 feet
_____	Texas Commerce Tower, Houston	1,000 feet
_____	World Trade Center, South, New York	1,362 feet
_____	Amoco, Chicago	1,136 feet
_____	First Interstate World Center, Los Angeles	1,018 feet
_____	World Trade Center, North, New York	1,368 feet
_____	John Hancock, Chicago	1,127 feet

1. The Sears Tower in Chicago, Illinois, became the world's tallest building when it was built in 1974. This beautiful building climbs 110 stories into the sky. How many feet per story does the building average rounded off to the nearest foot? ____________

2. Comparing height in miles, which of the following fractions of a mile comes closest to estimating the height of the Sears Tower?
   1/10 mile   1/8 mile   1/4 mile   2/5 mile   1/2 mile   3/4 mile

3. Standing on the observation deck at the Sears Tower, how much lower (in feet) would you look to see the second tallest building in Chicago's skyline? ____________

   ____

4. The Empire State Building, constructed at 34th Street and 5th Avenue in 1931 in New York City, was for many years the tallest building in the world. In 1973 the World Trade Center was completed and replaced the Empire State Building as the world's tallest building. How much taller is the World Trade Center (North) than the Empire State Building? ____________

5. The CN Tower in Toronto, Ontario, is the world's tallest self-supporting structure rising to a total height of 1,821 feet! How does that size compare with that of the Sears Tower in Chicago? ____________________________

6. The Stratosphere Tower in Las Vegas, Nevada, was completed in 1996 and billed as the "tallest free-standing observation tower in the United States" as well as the "highest structure west of the Mississippi." Its total height is 1,149 feet. How does the Stratosphere Tower compare to the largest skyscraper west of the Mississippi (the First Interstate World Center in Los Angeles)? ______________________

7. If you converted the height of the Sears Tower to meters, how tall is this majestic skyscraper in meters to the nearest whole meter? ___________

8. How many football fields "tall" is the Empire State Building? Round to the closest entire field. ______________________________

9. Which of these percents of a mile comes closest to estimating the height of the Texas Commerce Tower in Houston?
   15% 17% 19% 22% 26% 30%

10. If the Texas Commerce Tower in Houston has floors that are 13.33 feet in height, how many stories are there in the building? ___________

11. The Amoco Building in Chicago has floors that are 14.2 feet in height. How many floors are there in the entire building? ___________

12. Looking at the chart on the first page should quickly suggest to you that most of the nation's tallest skyscrapers are located east of the Mississippi River. Can you think of logical reasons for there being more tall buildings in the East than in the West?

____

____

Name ________________________

Mother Knows Best

Amy and Dave Bueschel have three children, Joe, Adam, and Betsy. Since the day each child was born, Amy has kept "growth charts" on her children. She carefully records their height and weight every six months as well as recording when each child acquires certain skills and proficiencies.

Amy decided to let her children know what a genuine Math Whiz she really is, so she told them that she could tell how much each child weighed without knowing the individual weight of any one of them. Her kids were really impressed, but they didn't believe her because they had never considered Mom to be a Math Wizard. This is what Amy did.

First . . . she asked Adam and Joe to get onto the scale together. When they did this, the scale recorded 216 pounds.

Next she asked Betsy and Joe to get onto the scale together. This time the scale recorded a weight of 185 pounds.

Finally, Amy asked Betsy and Adam to get onto the scale together. When they did, the scale recorded 173 pounds.

1. What strategy did you use to figure out how much each child weighed?

   ____

   ____

2. What would the scale have read if Amy had asked all three of her children to get onto the scale at the same time? ____________

3. How much did each child weigh?

   Joe ____________

   Adam ____________

   Betsy ____________

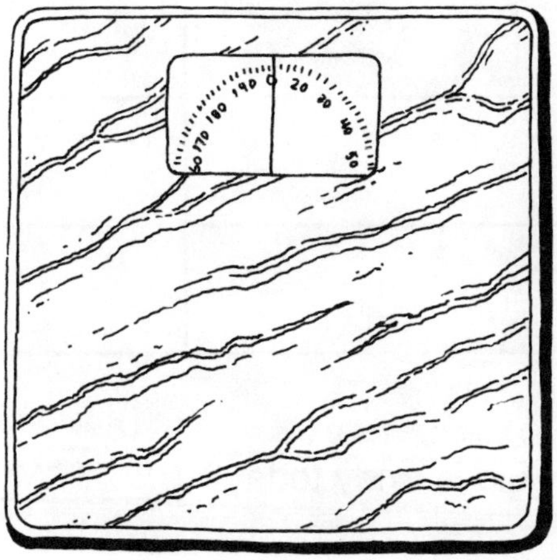

Happy Birthday, Andy

Andy is celebrating a birthday today. To mark the occasion, Andy's sister Gretchen went shopping with her friend Andrea. While they were shopping in the mall, Andrea asked Gretchen which birthday Andy was celebrating.

Gretchen thought about it for a moment, then replied with this information: "Andy's age today plus Dad's age total 60 years. Mom's age and Andy's age together add up to 54 years. And Mom's and Dad's ages together total 90 years."

Use the chart below to record everything you know and help Andrea solve the birthday puzzle of Andy's age.

Andy	Dad	Mom	Total Years
			60
			54
			90
			114
			24
			12
			48
			42

How old is Andy today? ________________

Name ______________________

Tickets, Please

People were lining up waiting impatiently for the Cinema 16 to open. At precisely 1:30 p.m. the ticket window opened. The first person stepped forward to announce that he wanted to pay for one adult and three student admissions to Theater #3. The ticket seller responded with, "That'll be $18.00."

The next person followed with this statement, "I'll need two adults and two children for Theater #7." The ticket seller responded with, "That'll be $21.00." Based on this information, find out the cost of an adult and a student admission. Then fill in the missing information on the chart.

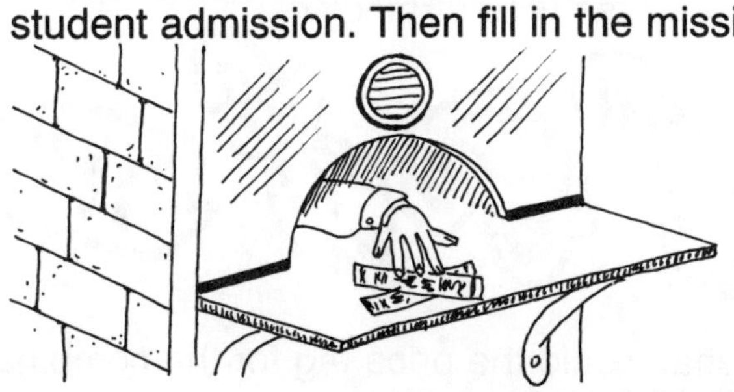

Cost of child admission ____________

Cost of adult admission ____________

Children \ Adults	0	1	2	3	4	5
5						
4						
3		$8.00				
2			$21.00			
1						
0						

Name ______________________________

Shades of Summer

We all know that it makes good sense to protect our bodies from too many harmful rays from the sun. When we go to the beach, it is wise to apply a lotion containing a sun block. We also use sunglasses to protect our eyes and umbrellas to provide shade at the beach. Soak up the prices on these "shades of summer" and answer the questions that follow.

1. Just by looking at the two picture combinations, which is the more expensive, a pair of sunglasses or the umbrella?
   What was your strategy?

2. What would the price tag for the combination below read?

3. How much for this combination?

4. How much for this combination?

5. Armed with this information, develop a strategy to figure out what each item costs. Explain your strategy on the back of this page.

Sunglasses 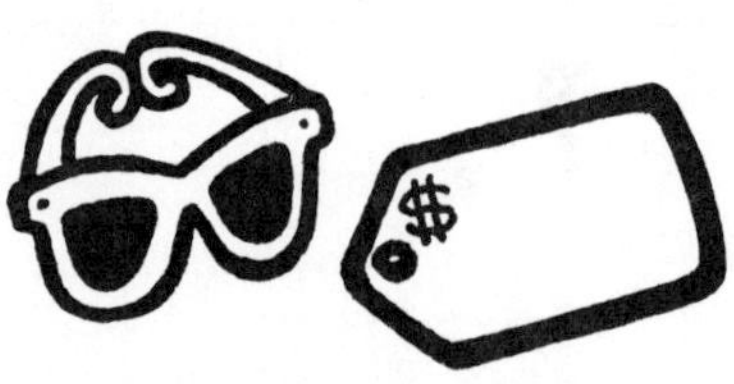

Umbrella 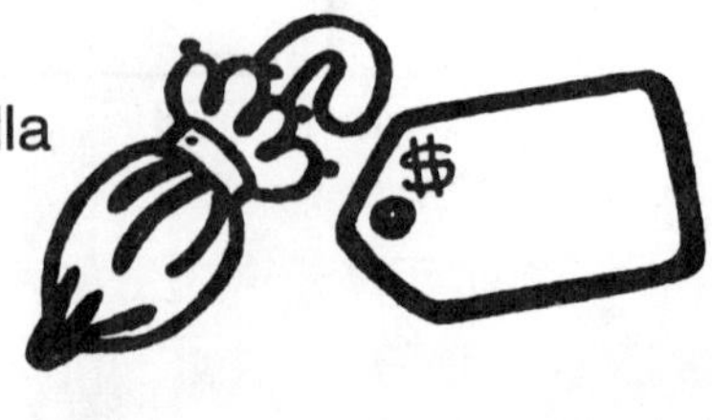

Name ______________________________

It's About Time

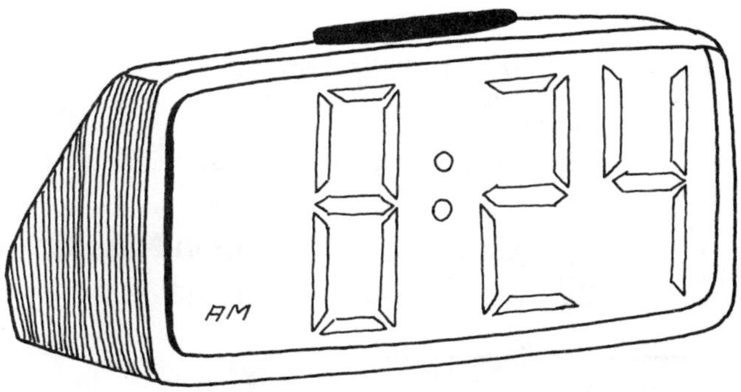

Changing time zones in the United States can be tricky business! The Continental United States spans four different zones resulting in the necessity for people to mentally calculate the difference in time when dealing with people living or working in different areas of the country. To some, this is a somewhat formidable task, but it is really quite simple if you take the time to understand.

Actually, it all has to do with the rotation of the earth. The earth rotates on its axis in a west-to-east direction, moving constantly *toward the sun.* Each day begins with the sun "rising" in the east and "setting" in the west. Planet Earth is divided into 24 time zones. Each day is born on the International Date Line, which is an imaginary line in the Pacific Ocean that extends from the North Pole to the South Pole. Its location was chosen because it does not cut through major bodies of land. It can be very confusing because crossing the Date Line can advance you an entire day (if you cross the line to the west) or lose an entire 24-hour day (if you cross the line to the east).

The reason for the various time zones is to give meaning to the time-of-day that will be common to everyone. When it is 4:00 a.m., you would expect people to be enjoying the last few hours of sleep during a given night. If it were the same time everywhere on the planet, there would be places where 4:00 a.m. would mean the sun was at its peak for the day and others where it was just beginning to set. Dividing the earth into 24 time zones places people in different areas of the earth into different time segments of the day, but it gives them all a sense of "the same" based on when they experience the various stages of daylight and darkness.

The map that accompanies this assignment shows you the dividing lines between the four time zones experienced by Americans in the continental United States. From that you can calculate the problems below. For example, if it is 4:00 p.m. in New York City, you can easily see that it is 3:00 p.m. in Chicago. Also if it is 2:30 a.m. in Los Angeles, then it is 3:30 a.m. in Denver. Use the same kind of logic in solving the problems below.

1. A major league baseball game between the New York Yankees and the Chicago White Sox is being played in Chicago. Game time is scheduled for 7:05 p.m. What time should a Yankee fan living in Burlington, Vermont, tune in to make certain none of the game is missed? ___________

2. If it is 3:15 p.m. in Denver, what time is it in Chicago? ______________

3. If it is 6:00 a.m. in Philadelphia, what time is it in Seattle? ______________

4. The difference between the time in Kansas City and San Diego is ______________

5. What time is it in Las Vegas when it is 3:00 a.m. in Boston? ______________

6. If it is 12:00 midnight in Pittsburgh, what time is it in Baltimore? ______________

7. When it is 5:00 a.m. in St. Louis, what time is it in Portland? ______________

8. A parade is scheduled to begin in New Orleans at 7:30 p.m. If someone in San Diego wants to watch the parade on television, what time should that person "tune in"?

____

9. The various time zones are distinguished from one another by the letters representing each zone. The letters EST represent Eastern Standard Time, CST means Central Standard Time, MST stands for Mountain Standard Time, and PST means Pacific Standard Time. Bearing this in mind, indicate the following differences in time zones. Be certain to tell whether the clock should be advanced or set back in each case.

PST to EST ______________

CST to PST ______________

EST to MST ______________

MST to PST ______________

MST to EST ______________

CST to EST ______________

PST to CST ______________

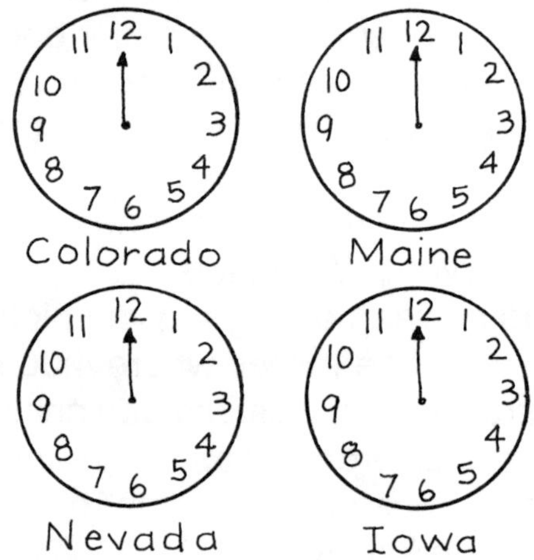

10. If Gretchen (who lives in Albuquerque) wants to call her aunt Polly (who lives in Buffalo), what time should Gretchen place the call if she wants to wake her aunt at 6:45 a.m. to wish her a happy birthday?

____

11. Matt, who lives in San Francisco, wants to call his grandma, who lives in Indianapolis. He wants to call her an hour before her 9:00 p.m. bedtime. What time should he call?

____

12. A flight leaves St. Louis at 7:20 a.m. and is scheduled to arrive in New Orleans one hour and forty-five minutes later. What time is the flight scheduled to arrive in New Orleans? ______________

13. A plane leaves Phoenix at 9:08 a.m. The flight time between Phoenix and Portland is two hours and eighteen minutes. What time is the plane scheduled to land in Portland? ______________

14. A flight leaves Chicago at 8:05 p.m. CST for a flight to Atlanta that lasts one hour and forty-two minutes. If the flight is on time, what time does the passenger arrive in Atlanta? ______________

15. Beginning the first Sunday in April and continuing through till the last Sunday in October, almost all of the United States experiences what is called "Daylight Savings Time," when the clocks are advanced an entire hour to give people more daylight during the evening hours of summer time. An exception to this rule is Arizona, which remains on Mountain Standard Time. If a plane leaves Tucson at 8:46 a.m, what time does it arrive in Chicago if the flight time is two hours and twenty-one minutes?

    ____

16. Memphis has a current time of 7:09 p.m. A plane leaves the airport on time for a flight to Salt Lake City. If the plane lands at 8:42 p.m, how long is the flight?

    ____

17. A plane leaves LaGuardia Airport in New York City at 4:56 p.m. The flight to Chicago lasts two hours and fourteen minutes. What time does the plane arrive at O'Hare in Chicago if it is eighteen minutes late? ______________

18. American Flight 708 leaves Phoenix at 7:09 a.m. on July 1. It has a scheduled flying time of two hours and thirty-one minutes to Detroit. If the plane arrives fourteen minutes behind schedule, what time is it when the passengers leave the plane?

    ____

19. United Flight 1208 leaves Philadelphia at 9:00 a.m. The two hours and twelve minute flying time gets it to Chicago by its scheduled arrival. In Chicago it refuels and restores food and drink during its forty-nine minute stop. Someone on Flight 1208 continues on to San Francisco, where the plane lands on time two hours and thirty-six minutes later. What time does the plane land in San Francisco? ______________

20. On the back of this paper, make up your own time zone problem and exchange it with a friend. Check each other's problems and then compare your answers.

    ____

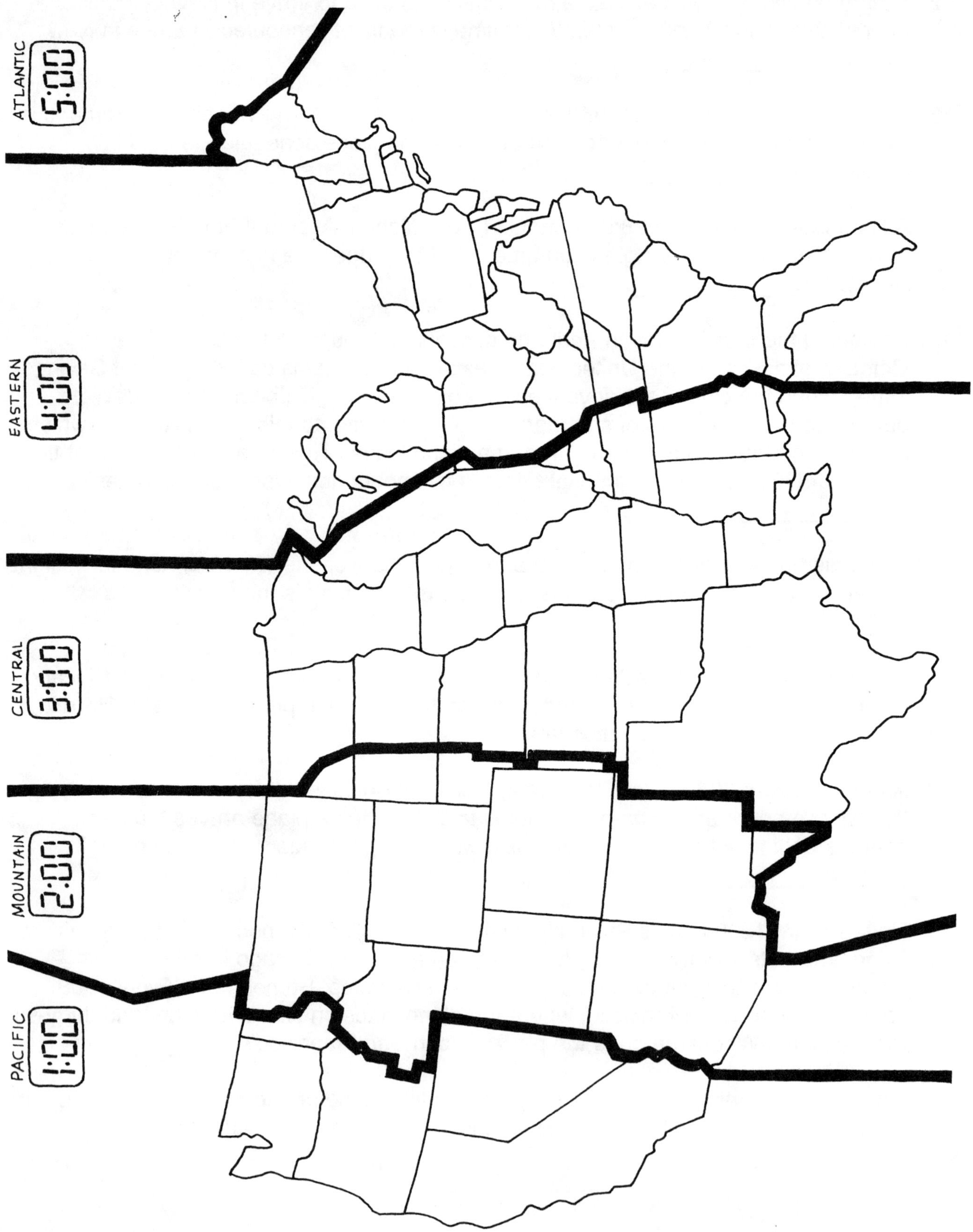
ATLANTIC
5:00
EASTERN
4:00
CENTRAL
3:00
MOUNTAIN
2:00
PACIFIC
1:00

Name ____________________

What Are My Chances?

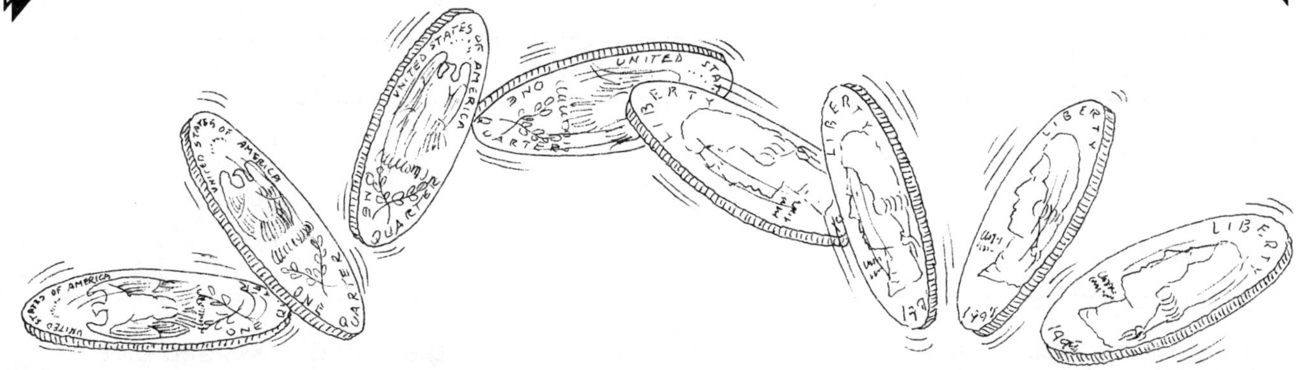

We all know that a coin has two sides and that if we flip the coin, the chances of that coin landing with a "heads" showing are the same as with a "tails" showing. At some point every middle school student has experienced a coin flip to determine who gets the biggest piece of cake, who gets to "go first," or who gets his/her choice. You probably never quibbled about a coin toss because you knew that it was a fair way of determining an outcome between two people.

If you had called "heads" and the coin landed with "heads" showing, what would have been your chances of winning the next time the coin was flipped if you stayed with your choice of "heads"? The answers to these and other questions will become clear as you work through these problems on probability.

The likelihood of an event happening is often written as a fraction that has been reduced to its lowest terms. For example, when the event is the toss of a coin, the chances of winning are one in two. That can be expressed as 1/2. Another way to express this relationship is 1:2.

1. You need a partner and a coin for this part of the assignment. If you were to flip coins with your partner 50 times, how many times would you expect to win? ________ How many times should your partner win? ________ Actually flip the coin 50 times. How many times did each of you win? Your number of wins ________ Your partner's number of wins ________ Did you both win exactly the same number of times? ________ If one of you had called "heads" on the first flip and won, what do you think your chances would have been of winning if you called "heads" again?

2. In a 52-card deck of cards, what are your chances of drawing an ace on the first draw? Express your answer as a fraction. ____________

3. If you did draw an ace, and did not replace it in the remaining deck, what are your chances of drawing another one when you draw again? ________________

4. Again, you will need a partner for this part of the activity. You will also need a pair of dice. In the chart below, calculate the various combinations that can be rolled to lead to each sum indicated. For example, the number 2 could only have one combination, that being a 1 and another 1. Indicate each of the other possible combinations in the space provided.

2	3	4	5	6	7	8	9	10	11	12
1,1										

How many total combinations are possible when throwing a pair of six-sided dice? ________ Simply add up the combinations you calculated above.

Now express the odds on rolling any given total on a single roll of the dice. Again, simply count the possibilities and express them as a fraction.

2 = 1/36 or 1:36   3 =   4 =   5 =

6 =   7 =   8 =   9 =

1   0 =   11 =   12 =

Now you and your partner should "test" the odds. Take turns rolling the dice and record your totals. Roll the dice18 times each or a total of 36 times. How many times did each sum appear?

2 =   3 =   4 =   5 =   6 =

7 =   8 =   9 =   10 =   11 =   12 =

How do these "real world" totals compare with your calculated odds?

____

5. Look at the spinners below. What are your chances of landing on each number indicated beneath each spinner? Indicate each answer as a fraction.

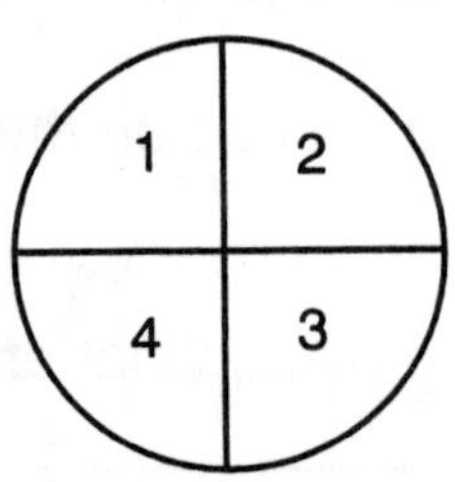

2 = ________

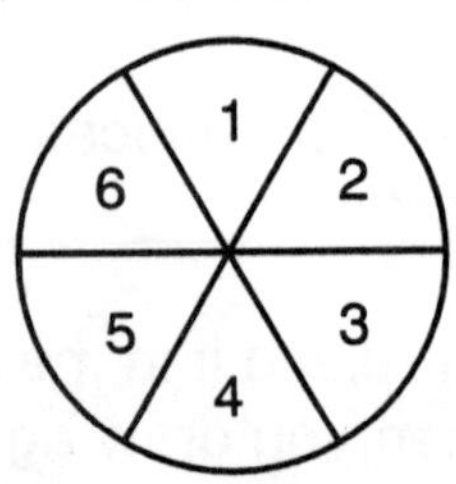

5 = ________

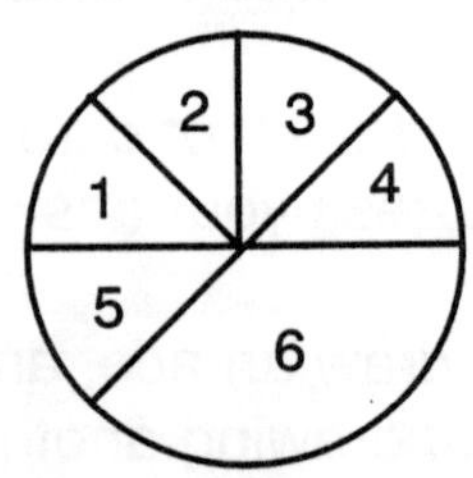

6 = ________

ANSWER KEY

Money, Money, Money

Face Value **pages 4-5**

penny	1¢	Lincoln	16th U.S. President
nickel	5¢	Jefferson	3rd U.S. President
dime	10¢	Roosevelt	32nd U.S. President
quarter	25¢	Washington	1st U.S. President
$1 bill	$1.00	Washington	1st U.S. President
$2 bill	$2.00	Jefferson	3rd U.S. President
$5 bill	$5.00	Lincoln	16th U.S. President
$10 bill	$10.00	Hamilton	1st U.S. Treasury Sec.
$20 bill	$20.00	Jackson	7th U.S. President
$50 bill	$50.00	Grant	18th U.S. President
$100 bill	$100.00	Ben Franklin	famous patriot and inventor

$500 bill—William McKinley
$1,000 bill—Grover Cleveland
$10,000 bill—Salmon Chase
$100,000 bill—Woodrow Wilson

Megabucks **page 6**

$1 bills	$5,674,996,280
$2 bills	$973,910,720
$5 bills	$6,672,449,500
$10 bills	$13,205,056,280
$20 bills	$75,061,576,480
$50 bills	$41,046,149,300
$100 bills	$207,539,210,500
Total	$350,173,349,060

1. $1 bills
2. $2 bills
3. $100 bills

Foreign Currency **pages 7-8**

1. U
2. C
3. L
4. F, DD
5. O, CC, JJ, GG
6. Q
7. N
8. X, FF
9. I
10. E
11. K, HH
12. A
13. R, S
14. T
15. V, Y, Z
16. P
17. B, MM, KK
18. D
19. M
20. J
21. W, AA
22. II, NN
23. G, EE
24. H

Big Budgets—Big Debts! **pages 9-10**

1991 $221,000,000,000
1992 $290,000,000,000
1993 $255,000,000,000
1994 $235,000,000,000

1. 1992
2. 1991
3. $1,001,000,000,000
4. $15,846
5. Answers will vary according to year.
6. Answers will vary.
7. Answers will vary.

Where the Money Comes From . . . Where It Goes **pages 11-13**

1. income taxes on individuals
2. insurance, hospital, and retirement taxes
3. 8 percent
4. $462 billion
   $387 billion
   $212 billion
   $100 billion
   50 billion
   37 billion
5. Answers will vary.
6. Answers will vary.

Checking It Out! **pages 16-20**

Check Register

Check No.	Date	Description of Transaction	Payment	Deposit	Balance
					493.61
1234	06/03	Smitty's	28.08		465.53
1235	06/04	Air Touch Paging	32.33		433.20
1236	06/04	Magpie's	18.45		414.75
	06/05	Deposit Janet's check		740.00	1,154.75
1237	06/05	Chevron	76.15		1,078.60
1238	06/06	Augat	95.00		983.60
1239	06/07	Foothills	40.00		943.60
1240	06/07	Shell Oil	62.27		881.33
	06/09	Deposit Tim		1,200.00	2,081.33
1241	06/09	Wells Fargo	501.40		1,579.93
1242	06/09	Canada Hills	25.16		1,554.77
1243	06/09	Catilina Power Company	147.88		1,406.89
	06/10	Withdrawal ATM	60.00		1,346.89
1244	06/10	Fry's	75.16		1,271.73

Plastic Peril! **pages 22-24**

1. $3,256.60
2. October 31, 1996
3. November 18, 1996
4. $97.70
5. 5246–8002–0421–0628
6. $1,743.40
7. They subtracted the charged purchases from the total credit line.
8. call customer service at 1–800–904–8357
9. 3 percent
10. $1,814.52
11. $47.39
12. $5,020.81
13. $196.60
14. Answers will vary, but most students will see the problems involved of overspending of credit cards.

We the People

Everyone Counts **pages 28-29**

1. White—199,714,000
   Black—30,094,000
   Hispanic—22,384,000
   Asian—7,213,000
   American Indian—1,990,000
2. Pie graphs will vary.
3. Mexican

Biggest Cities on Earth **pages 30-31**

1. Japan
2. Mexico
3. Brazil
4. Korea
5. United States
6. Japan
7. India
8. India
9. Brazil
10. Argentina

1. 25,018 people/square mile
2. Mexico City has 15,018 more people/square mile
3. 127,379 people/square mile
4. Hong Kong has over 120,000 more people/square mile
5. 535 square miles
6. Calcutta—59,628
7. 342 square miles
8. 11,480 people/square mile

Big City, USA **pages 32**

1. Los Angeles, Houston, San Diego, Dallas, Phoenix, San Antonio
2. New York, Chicago, Philadelphia, Detroit
3. Los Angeles
4. Phoenix
5. Detroit
6. Detroit

Where We Live **pages 33-36**

1. 48.75
   51.25
2. Bar graphs will vary.
3. South and West
   Northeast—3.4 percent
   Midwest—1.4 percent
   South—13.4 percent
   West—22.3 percent
4. People are tending to drift to warmer, less crowded areas. There are more available jobs in these areas of expansion.
5. Nevada—50.1 percent
   Alaska—36.9 percent
   Arizona—34.8 percent
   Florida—32.7 percent
   California—25.7 percent
6. New Hampshire—20.5 percent
7. Minnesota—7.3 percent
8. West Virginia—8.0 percent
   Iowa—4.7 percent
   Wyoming—3.4 percent
   North Dakota—2.1 percent
9. Florida—32.7 percent
10. West Virginia
11. Alaska—36.9 percent
12. California
13. Wyoming
14. California, which experienced a dynamic growth between 1980–1990, is currently losing population because of high costs of living there and people losing jobs because of companies relocating where costs are less expensive.

The Graying of America **pages 37-38**

1. Many older Americans move to warmer areas (like Florida) when they retire.
2. CA, ID, UT, NM, TX, LA, MS, GA, AK
3. Many of the southern states have a lower median age which may mean more job opportunities. There are no general conclusions about location for where older median ages live.
4. children born after World War II years (i.e., 1946 to 1964), 36–50 years old
5. As baby boomers grow another ten years older, they will shift the age even higher.
6. -They need to know how many people will be eligible.
   -They will need to know if new schools need to be built, study class size, teacher needs, etc.
   -They will need to know how many people will be needing assistance to help in budget figures plus number of people needed to administer to programs.
   -Companies need to know where to gear their research.
7. Answers will vary.

Equal Representation **pages 39-41**

1. California—52
   New York—31
   Texas—30
   Florida—23
   Pennsylvania—21
2. Montana, Alaska, Wyoming, South Dakota, Delaware, Vermont, North Dakota
3. 571,747
4. Alaska, Wyoming, Vermont
5.

States Gaining	States Losing
#1 California	#1 Montana
#2 Washington	#2 Kansas
#3 Arizona	#3 Iowa
#4 Texas	#4 Illinois
#5 Virginia	#5 Michigan
#6 North Carolina	#6 Ohio
#7 Georgia	#7 Pennsylvania
#8 Florida	#8 New York
	#9 West Virginia
	#10 Kentucky
	#11 Louisiana
	#12 New Jersey
	#13 Massachusetts

6. 19
7. Answers will vary.

Metro Crime **pages 42-45**

1. violent crimes per thousand people and property crimes per thousand people
2. murder, rape, robbery, aggravated assault
3. burglary, auto theft, larceny
4. Miami, FL—Dade County
5. Miami, FL—Dade County
   Gainesville, FL
   Tucson, AZ
   Myrtle Beach, SC
   Tallahassee, FL
6. All of these cities are in the South.
7. Broward County, FL
   Baton Rouge, LA
   Palm Beach County, FL
   Jackson, MS
   Fresno, CA
8. All five are warm belt cities.
9. The study used metro area rather than inner city, so while these cities may be dangerous within the inner city, the number of people added to the entire metro population may tend to diminish the overall crime statistics.
10. * More crimes are always committed when the weather is warm, and all of the top ten cities are in warm weather areas.
    * Tourist areas suffer from high levels of theft.
    * Also tourist cities have many more people in them than the Census Bureau thinks, hence the numbers would reflect a higher crime rate than is actually the case.
11. Answers will vary.

Education Pays **pages 46-48**

1. 45 to 54
2. $18,379
3. $55,518
4. $27,210
5. 45 to 54
6. 45 to 54

7. $30,040
8. $11,745
9. $21,476
10. $28,264
11. 45 to 54
12. Answers will vary, but data in chart point clearly to fact that men make higher salaries than women.
13. $763,310
14. $1,891,710
15. $1,128,400
16. $763,300
17. $1,418,520
18. $655,220
19. Answers will vary—higher self esteem, more social opportunities.

Heating Our Homes **pages 49-50**

1. Natural Gas—46,920,000
   Electricity—23,920,000
   Fuel Oil—11,040,000
   Bottled Gas—5,520,000
   Wood, Coal, or Solar Energy—3,680,000
   Some Other Fuel or No Fuel—920,000
2. 57 percent
3. natural gas or electricity
4. Graph will vary.
5. Answers will vary.
6. There are many power plants throughout the South that can generate electricity less expensively than in some other parts of the United States.

Buy, Sell, Trade

Tale of the Tapes **pages 54-57**

1. Super K-mart $107.57
   Albertson's $108.99
   Safeway $109.95
   ABCO $111.27
   Smith's $112.89
   Smitty's $113.76
   Fry's $113.35
   Bashas' $114.37
2. Super K-mart
3. Bashas'
4. $1.42
5. $6.80
6. Smitty's and Fry's
7. Super K-mart, Smith's, $2.80
   It could have been a much higher quality of beef at Smith's.
8. Safeway and Bashas'
9. Bashas'
10. Fry's
11. ABCO, Smith's, Fry's, Bashas' all had milk for $1.89.
12. 6¢
13. 79¢
14. Safeway, Smitty's, Fry's, Bashas'
15. Because it is still $1.00 cheaper at ABCO any time.
16. Bashas'
17. none
18. Safeway
19. Albertson's and Safeway at $1.99 for 5 pounds
20. Super K-mart, Albertson's, Smittys, and Fry's all had 64 oz bottle for $3.59.
22. Convenience, special services, and location are all important to consumers.

Fast Food and Fat **pages 58-63**

1. 40 grams
2. 13
3. Roast Chicken Salad, 149, 2
4. 381, 35
5. 1/4 Chicken, white or dark, Turkey Sandwich, skinless Rotisserie Turkey Breast; Chicken Sandwich and Ham Turkey Club would not be bad choices.
6. 240 calories left, but have already extended far beyond 36 grams of fat
7. The salad is fine. It is the ranch dressing that is high in both calories and fat.
8. Colonel's chicken and baked beans
9. Sausage is very high in fat plus the deep dish style adds both calories and fat.
10. Answers will vary, but the edible shell it is served in would have a high fat content.
11. Veggies
12. Because of the additions of mayonnaise to both main ingredients
13. 563 calories—only 8 grams of fat left
14. Answers will vary according to student preference.
15. 646 calories—34 grams of fat
16. 140 more calories at McDonald's and 6 more grams of fat
17. Answers will vary.
18. 2,383 mgs of sodium
19. Answers will vary.
20. No answer needed.

Little Anthony's Diner **pages 64-67**

1. Tim and Rhonda $12.40
   Raol and Maria $10.40
2. Time $15.28
   Raol $13.28
3. Tim $16.20
   Raol $14.08
4. Tim $2.50
   Raol $2.00
5. $28.90
6. $30.63
7. $3.00
8. Answers will vary.

Tyler's Sports Cards **pages 68-70**

1. $45.00
2. $40.50
3. $43.03
4. 2
5. Dish and Swish 8
   Block Party 4
   Rim Rockers 2
   Jam City 0
6. Diamond Kings 6
   Spirit of the Game 3
   Heritage Series 1
   Press Proof 1
7. 3 boxes
8. $120.49
9. $175.00. Probably, but there would be no assurance that he would get all players in the set even if he buys three boxes.
10. $68.60
11. $11.40
12. $76.00

Laurie's Laredo pages 71-74

1. $24,972.00
2. $2,501.00
3. $22,871
4. $9,172.00
5. $7,371.00
6. $6,671.00
7. $483.65
8. $7,753.93

House for Sale pages 75-78

1. $587.20
2. $7,046.40
3. $6,400.00, $646.40, $79,353.60
4. $177.60
5. $73,728.00 difference
6. $552.80
   $412.80
   $12,384.00
7. $139,760 interest
8. $520.56
9. $78.10
10. Answers will vary.

A Fact of Life pages 79-82

1. $159.00
2. $258.00
3. $624.00
4. $225.00 more for smoker
5. 22 years older
6. $295.00
7. $129.00
8. $3,915.00
9. $3,841.88
10. Premium = Projected dividend – Guaranteed payment x 75% x number of thousands of dollars in policy.
11. $1,252.50

The Sports Page

Following Baseball's Best Hitters page 86-87

1. .400
2. 5
3. Cordova, Minn, .340, 10
   Alomar, Balt, .376, 2
   Belle, Clev, .343, 5
   Boggs, NY, .339, 8
   Giambi, Oak, .362, 3
   Thomas, Chi, .341, 7
   Knoblauch, Minn, .3392, 9
   O'Neill, NY, .377, 1
   Brosius, Oak, .351, 4
   Molitor, Minn, .343, 6
4. Piazza, LA, .349, 8
   McGriff, Atl, .354, 6
   Gwynn, SD, .378, 1
   Bagwell, Hou, .355, 5
   Williams, SF, .340, 10
   Grace, Chi, .356, 4
   Grudzielanek, Mtl, .364, 2
   Segui, Mtl, .341, 9
   Joyner, SD, .362, 3
   Rodriguez, Mtl, .353, 7
5. 175 hits
6. 549 at bats

It Pays to Play! pages 88-90

1. $60,912,693
2. $40,741
3. $31,096 more per game
4. $9,100 each trip to the plate
5. $52,189,370
6. $1,799,633
7. $322,157 per game
8. $295,864 per game
9. $1,206,921 more
10. $1,120,434 more in 1996
    2142.3 percent increase over 1976

Are They Worth It? pages 91-93

1. $635,684 per game
2. $21,189 per minute
3. $53,596 per minute
4. 78 players
5. Answers will vary, but students will probably respond with same notions about how really overpaid athletes are.
6. $196,970 per game
7. $1,970 per pitch

Those Amazing Bulls! pages 94-96

1. .878
2. .951
3. .805
4. .732, .146
5. .658, .220
6. 12
7. 54
8. .98
9. 2
10. .512
11.

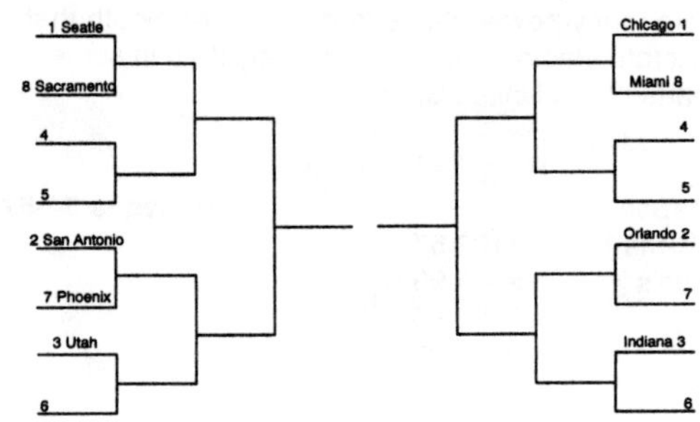

Predicting Winners pages 97-98

Answers will vary.

Batter Up! page 99

1. glove. Answers will vary, but "exchange" logic would suggest that a bat is less expensive than the glove.
2. $260.00
3. $260.00
4. $260.00
5. The glove cost twice as much as the bat.
6. 5
7. $26
8. $52

Big Time Franchise—Big Time Buck$ pages 100-101

1. 14 percent
2. $186 million
3. $153 million
4. 13 percent
5. 11 percent
6. $80 million
7. 11 percent
8. $174 million
9. $906 million
10. Warriors +23 percent
11. +7 percent

Take Me Out to the Crowd pages 102-103

1. $15,807,015.00
2. $204,020.78
3. 16,875.37 gallons
4. $283,055.85
5. $126,036
6. 4,999 fans

7. 50,053 T-shirts
   $1,101,166
8. 89,254,294 miles

Buy Me Some Peanuts and Crackerjacks! **page 104**
Hot dog $2.75
Crackerjacks $1.50
46-ounce soft drink $3.00
Peanuts $1.75

$uper Commercials **pages 105-106**
1. $40,000 per second
2. 33⅓ percent
3. 20 percent
4. $67.2 million
5. 28 minutes
6. 11.7 percent
7. $9.6 million
8. $12 million
9. $1.24 million
10. Answers will vary.

Torch of Peace **pages 107-109**
1. Maine, Vermont, New Hampshire, North Dakota, South Dakota, Montana
2. All six states are in somewhat remote locations with not a lot of people.
3. 84 days
4. 182 miles per day
5. 29 state capitals
6. 9,680 miles
7. 15 hours

The Fastest Man on Earth **pages 110-111**
1. 10.05 meters per second
2. 10.08 meters per second, he was moving .03 meters per second farther than Jim Hines per second
3. The 100-meter dash is 109.36 yards, 9.36 yards longer than the 100-yard dash. Accept various lines of thought; one is to compare distances in inches.
4. 33.06 feet per second
5. 2.75 feet per second faster
6. 22.54 mph
7. Lewis at 200 meters was traveling the fastest.
8. Answers will vary.
9. Answers will vary.

Marathon Man **pages 112-113**
1. approximately 22½ miles
2. 42,190 meters
3. 26 miles, 379 yards (actual marathons 26 miles 385 yards)
4. 12 miles per hour
5. 4 minutes, 56 seconds
6. 12.16 miles per hour
7. 10.9 miles per hour
8. 15 minutes, 31 seconds
9. Answers will vary.

Myself and Others

The Beat of Life **pages 117-118**
1. Estimates will vary.
2. a.. 4,200
   b. 100,800
   c. 705,600
   d. 36,691,200
   e. 2,751,840,000 times
3. Ten days is the most reasonable answer.
4. a. 1,755
   b. 3,300 more beats
5. a. 7,200 quarts
   b. 2,628,000 quarts
   c. 206,319,600 quarts
6. 4,065 quarts
7. Answers will vary.

Pound for Pound **pages 119-121**
1. 29
2. 72.5 percent
3. This study is far above the national average!
4. 5'6" 198 lbs 33
   6'0" 210 lbs 26
   5'9" 206 lbs 23
5. 6'2" 193 lbs 70
   6'1" 179 lbs 47
6. Younger people are considered more active and should weigh less.
7. Answers will vary.

Burning Calories **pages 122-124**
1. playing golf
2. jogging 11-minute mile
3. racquetball
4. They are very similar games requiring the same energy.
5. swimming the crawl stroke
6. running 8-minute miles
7. volleyball
8. 162 calories
9. 210 calories
10. 228 calories
11. 3,285 calories
12. 170 lb person burns 186 more calories
13. 200 lb person burns 384 more calories
14. 30 minutes of squash
15. one hour and 15 minutes of tennis
16. 66 more calories
17. 1,680 more calories
18. 300 more calories
19. jogger burns 9 more calories
20. golfer burns 69 more calories
21. Answers will vary.
22. Answers will vary.

Battle of the Math Classes **pages 127-128**
1. 124 feet, 5 inches
2. 125 feet, 3 inches
3. Mrs Little's class is 10 inches taller.
4. 249 feet, 8 inches
5. Answers will vary.
6. 5 feet, 2¼ inches
7. 5 feet, 2⅝ inches

Higher Learning—Higher Costs **pages 135-136**
1. $2,396.37 per year
2. $114,249
3. $12,058 per year
4. Answers will vary.

Math Et Cetera

Our Tallest Skyscrapers **pages 139-140**
4, 8, 7, 1, 10, 3, 5, 9, 2, 6
1. 13 feet per story
2. ¼ mile
3. Sears Tower is 318 feet taller than the Amoco Building.
4. 118 feet taller
5. CN Tower is 367 feet taller.
6. Stratosphere Tower is 131 feet higher.
7. 443 meters

8. Four football fields straight up would be 1,200 feet tall.
9. 19 percent
10. 75 stories
11. 80 floors
12. East has always been more densely populated which means space is more of a premium, hence, some of the early skyscrapers.

Mother Knows Best **page 141**

1. Answers will vary.
2. 287 pounds
3. Joe, 114 pounds
   Adam, 102 pounds
   Betsy, 71 pounds

Happy Birthday, Andy **page 142**

Andy	Dad	Mom	Total Years
X	X		60
X		X	54
	X	X	90
2	X	X	114
2			24
X			12
	X		48
		X	42

12

Tickets, Please **page 143**

Children \ Adults	0	1	2	3	4	5
5	$18.75	$25.50	$32.25	$39.00	$45.78	$52.50
4	$15.00	$21.75	$28.50	$35.25	$42.00	$48.75
3	$18.75	$8.00	$24.75	$31.50	$38.25	$45.00
2	$7.50	$14.25	$21.00	$27.75	$34.50	$41.25
1	$3.75	$10.50	$17.25	$24.00	$30.75	$37.50
0		$6.75	$13.50	$20.25	$27.00	$33.75

child admission, $3.75
adult admission, $6.75

Shades of Summer **page 144**

1. umbrella
2. $216
3. $204
4. $228
5. Sunglasses $30
   Umbrella $42

It's About Time **pages 145-148**

1. 8:05 p.m.
2. 4:15 p.m.
3. 3:00 a.m.
4. 2 hours
5. 12:00 a.m.
6. 12:00 midnight
7. 3:00 a.m.
8. 5:30 p.m.
9. advance 3 hours
   set back 2 hours
   set back 2 hours
   set back 1 hour
   advance 2 hours
   advance 1 hour
   advance 2 hours
10. 4:45 a.m.
11. 6:00 p.m.
12. 9:05 a.m.
13. 10:26 a.m.
14. 10:47 p.m.
15. 1:07 p.m.
16. 2 hours 33 minutes
17. 6:28 p.m.
18. 12:54 p.m.
19. 11:37 a.m.

What Are My Chances? **pages 149-150**

1. 25, 25
2. $\frac{4}{52}$ or $\frac{1}{13}$
3. $\frac{3}{51}$ or $\frac{1}{17}$
4.

2	3	4	5	6	7	8	9	10	11	12
1,1	1,2 2,1	2,2 1,3 3,1	2,3 3,2 4,1 1,4	3,3 5,1 1,5 4,2 2,4	4,3 1,6 3,4 6,1 5,2 2,5	4,4 5,3 3,5 2,6 6,2	5,4 4,5 6,3 3,6	5,5 4,6 6,4	6,5 5,6	6,6

There are 36 combinations possible.

$\frac{1}{36}$	$\frac{1}{18}$	$\frac{1}{12}$	$\frac{1}{9}$
$\frac{5}{36}$	$\frac{1}{6}$	$\frac{5}{36}$	$\frac{1}{9}$
impossible	impossible	$\frac{1}{18}$	$\frac{1}{36}$

5. $\frac{1}{4}$, $\frac{1}{6}$, $\frac{3}{8}$